Light-Based Science:
Technology and
Sustainable Development

Light-Based Science: Technology and Sustainable Development
The Legacy of Ibn al-Haytham

Edited by
Azzedine Boudrioua
Roshdi Rashed
and
Vasudevan Lakshminarayanan

CRC Press
Taylor & Francis Group
Boca Raton London New York

CRC Press is an imprint of the
Taylor & Francis Group, an **informa** business

CRC Press
Taylor & Francis Group
6000 Broken Sound Parkway NW, Suite 300
Boca Raton, FL 33487-2742

First issued in paperback 2019

© 2018 by Taylor & Francis Group, LLC
CRC Press is an imprint of Taylor & Francis Group, an Informa business

No claim to original U.S. Government works

ISBN-13: 978-1-4987-7938-8 (hbk)
ISBN-13: 978-0-367-88958-2 (pbk)

Library of Congress Cataloging-in-Publication Data

Names: Boudrioua, Azzedine, editor. | Råashid, Rushdåi, editor. | Lakshminarayanan, Vasudevan, editor.
Title: Light-based science : technology and sustainable development / edited by Azzedine Boudrioua, Roshdi Rashed and Vasudevan Lakshminarayanan.
Description: Boca Raton : CRC Press, [2018] | Includes bibliographical references and index.
Identifiers: LCCN 2017007536| ISBN 9781498779388 (hardback : alk. paper) | ISBN 9781498779401 (ebook)
Subjects: LCSH: Optical engineering. | Optics--Research--Middle East. | Solar power. | Light--Philosophy. | Sustainable development.
Classification: LCC TA1520 .L55 2018 | DDC 621.36--dc23
LC record available at https://lccn.loc.gov/2017007536

Visit the Taylor & Francis Web site at
http://www.taylorandfrancis.com

and the CRC Press Web site at
http://www.crcpress.com

Whoever seeks the truth is not one who studies the writings of the ancients and, following his natural disposition, puts his trust in them, but rather the one who suspects his faith in them and questions what he gathers from them, the one who submits to argument and demonstration and not the sayings of human beings whose nature is fraught with all kinds of imperfection and deficiency.

Ibn al-Haytham

Contents

Foreword and Acknowledgement..xi
Preface...xiii
About the Editors xv
Contributors xvii
Introduction...xix

SECTION I   History Guiding the Future, the Ibn al-Haytham Legacy

Chapter 1   Ibn al-Haytham's Scientific Research Programme 3

Roshdi Rashed

Chapter 2   Light and Vision before Ibn al-Haytham: The Perspective of
Islamic Theology...9

J Van Ess

Chapter 3   From the Reasons of Light to the Lights of Reason: Remarks on
the Nine Centuries Distant Ibn al-Haytham's and Albert Einstein's
Respective Approaches of Light as Conceived Physically 23

Michel Paty

Chapter 4   Translating and Interpreting Ibn al-Haytham's *Optics* from
Arabic to Latin: New Light on the Vocabulary of Reflection and
Refraction 43

Paul Pietquin

Chapter 5   Ibn al-Haytham: The Founder of Scientific Pluralism.......................53

Hassan Tahiri

Chapter 6   Ibn al-Haytham: Founder of Physiological Optics? 63

Vasudevan Lakshminarayanan

Chapter 7   Ibn al-Haytham's Problem...109

Pierre Coullet and Roshdi Rashed

Chapter 8   Ibn al-Haytham and His Influence on Post-Mediaeval Western
               Culture 123

               Charles M Falco

SECTION II   Light-Based Technologies for the Future

Chapter 9   Photonic Technology: Recent Developments and Challenges of
               the Twenty-First Century 135

               Azzedine Boudrioua

Chapter 10 Ibn al-Haytham's Thousand-Year Journey from Basra to Mars 143

               Noureddine Melikechi

Chapter 11 Ibn al-Haytham and the International Year of Light: His Legacy ... 151

               Mohamed M El-Gomati

Chapter 12 New Short-Wavelength Pulsed Light Sources 165

               Majed Chergui

Chapter 13 Lighting: From Human Evolution to Sustainable Revolution 181

               Harry Verhaar

Chapter 14 Mediaeval Arab Achievements in Optics 193

               Sameen Ahmed Khan

SECTION III   Optics and Photonics in the Arab and Islamic World, Education and Investment in Light Sciences and Technology

Chapter 15 Need to Create International Science Centres in Arab Countries ... 207

               Azher Majid Siddiqui and Sameen Ahmed Khan

Chapter 16 SESAME: The First Synchrotron Light Source in the Middle
               East and Neighbouring Regions 221

               Gihan Kamel

Chapter 17 Scientific Translation: A Tool from Shadow to Light 229

Abdulaziz Alswailem

Chapter 18 Optics and Photonics Research in Lebanon: Key Figures, Ways Forward 243

Marie Abboud Mehanna

Index...251

Foreword and Acknowledgement

This book is an important milestone of the work undertaken by the Ibn al-Haytham International Working Group (IWG) during the year 2015 to celebrate the International Year of Light and Light-based Technology (IYL2015). It contains the main contributions presented during the international conference organised at the UNESCO HQ in Paris in September 2015 and is dedicated to the legacy of Ibn al-Haytham as the precursor and the founder of modern science in general and optics in particular.

This conference would not have been possible without the support and contribution of His Excellency, the Ambassador of Saudi Arabia to the UNESCO, Dr. Ziad Aldrees. In fact, it all began with the important work of His Excellency in lobbying for the inclusion of Ibn al-Haytham in the resolution of the United Nations declaring IYL2015. Indeed, for the first time, the United Nations explicitly emphasised the coincidence of such an event with the immense contribution of Ibn al-Haytham to light science. More than that, from the first hour, His Excellency immediately supported the creation of the IWG group and the organisation of the conference.

We express our gratitude and acknowledge the encouragement and support of the Ambassador of the Kingdom of Saudi Arabia to the UNESCO.

We would also like to thank Professor John Dudley for his efforts in forming the IWG and making it active. We also express our warm thanks to Dr. Maciej Nalecz, former director of the Division of Science Policy and Capacity Building, and his wonderful team at UNESCO and, in particular, to Dr. Jean Paul Ngom Abiaga and Dr. Ossman Benchikh for the organisation of this conference and for supporting the IWG group.

Our sincere gratitude goes to all authors for contributing their chapters in a timely manner. This book would not exist without their contributions. We would also like to thank CRC Press/Taylor & Francis for swiftly agreeing to publish this work when the idea was proposed to them.

This collective work shows the spirit that we wish to give to our future activities, through the Ibn al-Haytham LHiSA international society, which is the successor of the IWG. The example of Ibn al-Haytham reached a very broad audience during 2015 and we have the responsibility to continue this important work.

This is the beginning of a human adventure that we hope is sustainable and, above all, benefits human culture and civilization.

Azzedine Boudrioua
Roshdi Rashed
Vasudevan Lakshminarayanan

Preface

People throughout the world and across history have always attached great importance to light. We see this in cultural symbolism, universal myths and legends, and in the many ways that studying the science of light and applying it in practical applications has shaped the societies in which we live. It was in recognition of the importance of light in so many areas of life that the United Nations proclaimed the year 2015 as *the International Year of Light and Light-based Technologies (IYL2015)*, and the year 2015 has seen unparalleled international cooperation to promote activities in outreach and education around the theme of light in its broadest sense.

We can be immensely proud of what we have achieved together. Brought together under the leadership of UNESCO, partners around the world organised an estimated 10,000 activities in over 140 countries. Although a number of these events were planned centrally by the Steering Committee and our recognised international partners, many activities were organised spontaneously on a local level all over the world.

A particular feature of IYL2015 has been the extremely broad support from all sectors. Although the original idea began back in 2009 among a partnership of scientific societies, it rapidly grew to include academic and industry organisations, NGOs, and it has even received support at the highest level with patronage from royalty in the United Kingdom and Spain, and events involving governments in many other countries. The range of events has been extremely diverse: education and outreach for students and the public; specialist workshops in science and industry; forums on the historical development of science; conferences on sustainable development; public light festivals and displays; works of art, music and literature. Events have been targeted at all levels – from preschool children learning science for the first time, to politicians and diplomats attending high-level meetings on the importance of technology for the future.

One of the particular themes of the year which was of central importance was celebrating the many works and influence of the pioneer Ibn al-Haytham. Events of all kinds under this banner took place worldwide, and a selection of papers delivered at a dedicated two-day conference at UNESCO headquarters in Paris are reproduced in this rich volume that you hold in your hands. The works of Ibn al-Haytham and the International Year of Light have inspired millions worldwide.

After the successes of the International Year of Light, the challenge for us now is to ensure that the partnerships made during 2015 will continue to develop. However, there is every reason to be extremely optimistic. The many different sectors involved in light science and its applications have shown what they can do when working together, and the foundations are now in place for a continued development that will build a better world for us all tomorrow.

John Dudley
Chairman of IYL2015

About the Editors

Azzedine Boudrioua is a full professor at the University Paris 13. He leads the Organic Photonics and Nanostructures group of the *Laboratoire de Physique des Lasers* (LPL) at Galilee Institute. Professor Boudrioua earned a PhD in physics from the University of Metz, France in 1996 where he worked as an associate professor for 15 years in the field of integrated optics and photonic crystals. In 1997, he joined the group of Professor W L Barnes at Exeter University (UK) as a fellowship researcher where he worked on microstructured luminescent polymers. Four years later (2001), he defended his *Habilitation à Diriger des Recherches* (HDR) at the University of Metz. Currently he is developing activities in the field of nano-photonics as well as nonlinear optics.

He actively participates in several national and international scientific committees and expert panels. He is involved in several projects at the national and international levels and leads some of them. He was also involved in the organization of several conferences. He is a member of the Research Council and the Restricted Academic Council of the University Paris 13 and president of the disciplinary commission.

He was a member of the Board of the French Optical Society (SFO) (2007–2015) and its treasurer between 2007 and 2012. He is the founder of several thematic clubs of the French Optical Society, including organic photonics and optical fiber and network clubs. He was a member of the International Steering Committee of the International Year of Light and Light-based Technologies (IYL) 2015 and he coordinated the Ibn al-Haytham International Working Group. He was recently elected as president of Ibn al-Haytham International LHiSA (Light, History, Science and Applications) Society for the period 2016–2020. Professor Boudrioua has published more than 96 journal papers and more than 140 papers were presented in national and international conferences. He also published two books *Integrated Optics* and *Organic Lasers*. He has received the Arnaulf-Françon Award in 2008 from SFO (French Optical Society).

Roshdi Rashed is the author of several books and scientific articles in History of Science. He is currently Emeritus Director of Research (special class) at CNRS (France). He was director of the Centre for History of Arab and Medieval Science and Philosophies (until 2001) Paris, and also director of the doctoral formation in epistemology and history of science, Paris Diderot University (until 2001). He is Emeritus Professor at Tokyo University, and at the Mansoura University, and also at the Paris Diderot University. He was a founder (1984) and director (until May 1993) of the Research Epistemology and History of Science and Scientific Institutions (REHSEIS) research team, CNRS, Paris. He had several distinctions including the CNRS Bronze medal (1977), Knight of the Honour Legion (1989), the Alexandre Koyré medal of the International Academy for History of Science (1990), the history of science medal and award of the Academy of Sciences for the Developing World (1990), medal and award of Kuwait Foundation for the Advancement of Sciences (1999), Avicenna gold medal of UNESCO (1999), medal of CNRS (2001), medal

of the Arab World Institute (2004). He had several honorary positions as vice president of the International Academy of the History of Science (1997), member of the
Royal Belgian Academy of Sciences (2002), member of the Tunisian Academy 'Beit
el-Hikma' (2012) and so on.

Vasudevan Lakshminarayanan (PhD University of California at Berkeley) is currently a professor of vision science, physics, electrical and computer engineering
and systems design engineering at the Unviersity of Waterloo. He has held visiting and faculty appointments at the universities of California at Berkeley, at Irvine,
University of Missouri, the University of Michigan, Glasgow Caledonian University
and the Indian Institute of Technology at Delhi and at Madras. He is also an adjunct
professor of electrical and computer engineering at Ryerson University (Toronto).
He has had numerous honours, including fellow of OSA, SPIE, AAAS, APS, IoP,
etc., a senior member of IEEE and the recipient of a number of awards including
most recently the Esther Hoffman Beller medal of OSA (2013), and the Optics
Educator award of SPIE (2011). He has been a KITP Scholar at the Kavli Institute
of Theoretical Physics at the University of California at Santa Barbara. He was
an AAAS science and technology policy fellow finalist and serves on the optics
advisory board of the Abdus Salam International Centre for Theoretical Physics in
Trieste, Italy. He is a founding member of the UNESCO ALOP programme. He also
served as member at large of the U.S. IUPAP committee of the NAS, the Chair of the
U.S. International Commission on Optics committee, Chair of the APS Committee
on International Scientific Affairs, a member of the international steering committee of the International Year of Light 2015, a member of the education committee
of the National Photonics Initiative and has served as a director of OSA. He is a
consultant to the ophthalmic and medical devices section of the FDA (from 2013).
He is technical editor or section editor of a number of journals including Journal
of Modern Optics (from 2003) and Optics Letters (from 2010). He is currently the
Chair of the education committee of SPIE (2015–2017). He has authored or edited
about 12 books, and over 300 publications in topics ranging from optical physics
and engineering, neuroscience, bioengineering, applied math and ophthalmology/
optometry.

Contributors

Abdulaziz Alswailem
King Abdulaziz City for Science and
    Technology
Riyadh, Saudi Arabia

Azzedine Boudrioua
Laboratoire de Physique des Lasers
University of Paris
Paris, France

Majed Chergui
Laboratoire de Spectroscopie Ultrarapide
Ecole Polytechnique Fédérale de
    Lausanne, ISIC

and

Lausanne Centre for Ultrafast Science
Faculté des Sciences de Base
Lausanne, Switzerland

Pierre Coullet
Université de Nice-Sophia Antipolis
Nice, France

Mohamed M El-Gomati
Department of Electronics
University of York
York

and

Foundation for Science
Technology and Civilisation
United Kingdom

Charles M Falco
University of Arizona
Tucson, Arizona

Gihan Kamel
IR beamline Scientists
SESAME
Jordan

Sameen Ahmed Khan
Department of Mathematics and
    Sciences
Dhofar University
Salalah, Sultanate of Oman

Vasudevan Lakshminarayanan
Depts. of Physics, Electrical and
    Computer Engineering and Systems
    Design Engineering
University of Waterloo
Waterloo, Ontario, Canada

Marie Abboud Mehanna
Physics Department
Saint-Joseph University
Beirut, Lebanon

Noureddine Melikechi
Department of Physics and Applied
    Physics
University of Massachusetts
Lowell, Massachusetts

Michel Paty
French National Centre for Scientific
    Research (CNRS)
Laboratory Sphere, CNRS
Université Paris
Diderot, France

Paul Pietquin
UR Mondes anciens
University of Liège
Belgium

Roshdi Rashed
Université Denis Diderot
Paris, France

Azher Majid Siddiqui
Department of Physics
Jamia Millia Islamia
New Delhi, India

Hassan Tahiri
University of Lisbon
Lisboa, Portugal

J Van Ess
Center for Islamic Theology
University of Tuebingen
Tuebingen, Germany

Harry Verhaar
The Netherlands

Introduction

It's Only a Story of Light …

In 2015, the world celebrated the International Year of Light and Light-based Technologies (IYL2015) declared by the United Nations. In the resolution (resolution A/RES/68/221 of 20 December 2013), the United Nations stressed the importance of light and light-based technologies in the lives of the citizens of the world and for the future development of human society. They also emphasised the coincidence of this event with the anniversary of several important stages in the history of light sciences including the foundation works of Ibn al-Haytham (1015), the wave theory of light from Fresnel (1815), the work of J C Maxwell in 1865, Einstein's theories on the photoelectric effect in 1905 and the links between light and cosmology revealed by the theory of general relativity in 1915, the discovery of the cosmic diffuse background by Penzias and Wilson as well as Kao's work on the transmission of light in fibres for optical communication in 1965. IYL2015 was a unique opportunity to recall and celebrate the contribution of Arab scholars over the past 1,000 years to human knowledge.

The success of Light science is above all the success of the universal knowledge built up over the centuries in which all civilizations and cultures have contributed to the edifice.

In ancient times, Light was considered an inner fire and a sense similar to touch, or even an extension of the soul to palpate and probe the outside world. In the Arab civilisation, light was perceived for the first time as a physical substance: the object of scientific study and understanding. The Arabs demonstrated interest in Light not only as a physical phenomenon but also studied seriously its behaviour such as the concentration of light by mirrors and lenses.

Today, Light and its myriad of applications continue to fascinate scientists in both the macroscopic and microscopic realms.

Light has become indispensable in policies of socioeconomic development in all sectors of human activity. It can be used to produce energy using sunlight, transmit information anywhere around the globe through fibre-optic cables, and it is used in medicine for both diagnostics and in therapeutic applications. Light is also a form of energy whose elementary particle, the photon, can interact with matter and give birth to physical, chemical and even mechanical phenomena.

The development of optics and photonics as a leading scientific and technological discipline took place during the last third of the twentieth century following the advent of the laser. The rapid development of photonics as a major scientific and technological endeavour is the result of a great deal of research and development, particularly in the field of guided optics, which has led to improved optical fibre performances and the development of optical telecommunications networks. Optical telecommunication networks are the main drive of photonic technologies. The research and development carried out in this field (guided optics) has made it possible to design low-cost optoelectronic components of all kinds. As a result, other applications in various fields have emerged.

Consequently, the use of Optics affects all aspects of the citizen's life: entertainment, medicine, energy and many other areas of everyday life and sectors as unsuspected as the automobile.

The beginnings of photonics (science of light) in the 1980s did not suggest the spectacular growth of this discipline. Photonics had remained limited to academic textbooks and university laboratories for decades. However, it has caught up quickly, notably by using the same technological tools developed by and for microelectronics. Today, we are considering integrated photonic circuits that can replace microelectronic circuits in the systems and components of tomorrow.

Replacing an electrical signal with an optical one which is 10,000 times faster will pave the way for many future applications and developments. This technological race is enabling photonics to emerge as the key technology of the twenty-first century and the next millennium. The photonics–electronics convergence that we know today heralds the advent of tomorrow's all-optical components and systems thanks, in particular, to Nanophotonics, which will allow to push the limits of miniaturisation of photonic systems to lower scales at the wavelength. As the physicist R Feynman predicted in 1959, we have entered the era of *Smaller, Faster and Cheaper*, which posited the possibility of manipulating matter at the atomic and molecular scale in order to conceive and to make sub-micrometric components and systems. In this nano-world, the photon is dethroning the electron and creating new opportunities.

This book is both the fruit and the logical continuation of the work carried out by the Ibn al-Haytham International Working Group (IWG) throughout the year 2015 and culminating with the international conference 'The Islamic Golden Age of Science for Today's Knowledge-based society: the Ibn al-Haytham example' at the headquarters of UNESCO on 14 and 15 September 2015. The conference focused on the life and works of Ibn al-Haytham, whose pioneering *Book of Optics* (*Kitāb al-Manāẓir*) was published approximately 1,000 years ago. It also emphasised the importance of linking the work of Ibn al-Haytham to key policy issues to address the challenges faced by society and to matters of importance to the general public (how light and light technologies offer solutions to many challenges and can positively impact people's daily lives and needs), particularly in developing countries. Eventually, the conference concluded with a highlight of the situation of research and education in Arabic and Islamic countries in the field of Optics and Photonics.

This conference combined the contributions of renowned historians and scientists working on light with the important one by Professor Roshdi Rashed who played a leading role in setting up the IWG group and in organising the conference. In fact, he is behind everything. He is our guide and our master.

The main objective of the IWG was to determine the best ways to celebrate and disseminate the contribution of Ibn al-Haytham and other Arab scholars to science in general and Optics in particular. A common international view of the contribution of Arab scholars was proposed and managed by the working group under the guidance of the international steering committee.

The IWG included leading Optics and Photonics scientists as well as historians and it is truly an international group with scholars from Africa, Europe, North America and Asia.

During the IYL2015, the IWG also proposed and organised specific sessions devoted to the Arabic scholars' contributions, which were included in local domestic programmes in various Islamic and Arab countries.

The launch of IWG began on 19 February 2014 with a fruitful discussion with Professor John Dudley, the chairman of the IYL2015 steering committee. We rapidly came to the decision to set up an international group of scholars.

This challenging issue was for me an exciting opportunity to finally fulfil an ambition of mine that I had for many years. In fact, when studying Physics in my native country of Algeria, I was fascinated by Western scholars. In general, studies of science in educational institutions (even in Algeria where I started studying Physics) bothered me because the discussions focused on the Greek heritage, leading directly to the Western scholars of the Renaissance and the age of Enlightenment. Later on, I realised that we need to rediscover the missing link between the ancient world and the European Renaissance period as well as the forgotten legacy of Islamic scholars.

What better example of a forgotten legacy than that of Ibn al-Haytham who enlightened the whole science of the Middle Ages and beyond. Ibn al-Haytham undertook the reform of the Optics as a builder who makes a masterpiece for posterity. This illustrious scientist has laid the foundations of modern Optics and experimental Physics. He developed the basis of our modern theory of vision and light. Ibn al-Haytham was the first to introduce the 'scientific method' (usually attributed to Francis Bacon), namely experimentation, observation and analysis which is very similar to the modern one.

IYL2015 has raised hope, enthusiasm and also curiosity to know much more not only about Ibn al-Haytham, but also about *The Islamic Golden Age of Science*. Personally, I was very touched by the expectations of young people during various meeting in high schools in France, particularly in the suburbs of Paris.

More than ever, the present-day events push us to continue the dissemination of the work of scholars of the Islamic civilisation such as Ibn al-Haytham. We must continue to highlight the importance of the contribution of Arab scholars to human civilisation in our education systems. It is important to teach our children, especially those who originate from Arab and Muslim countries, that they belong to an important civilisation and can contribute to human civilisation as their ancestors did.

This is particularly important and urgent in view of the general situation of the Arab and Muslim world as reported in the recent UNESCO report (UNESCO science report, towards 2030). The Arab world is of major importance (owing to its location and its wealth of oil and natural gas). Indeed, the economic slump due to the fall in oil prices has hit hard many countries of the region and profligate military spending is devouring the scarce resources for development.

We do need to focus on the following priorities: socially relevant problem-solving research, mobility and education and synchronising university curricula to market needs, as well as many other issues. The Network for the Expansion of Convergent Technologies in the Arab Region (NECTAR) programme is one of the actions that might bring a real scientific and technological development in the Arab world. As scientists, our contribution is of primary importance and we should play a central role in solving the problems faced by humanity such as health, energy, food, etc.

For all these reasons, our work during 2015 has motivated the creation of a new scientific international society. I am very delighted to announce the creation of the Ibn al-Haytham International Light: History, Science and Applications (LHiSA) Society). This is a non-profit society registered in France (www.ibnalhaytham-lhisa.com).

The Society's aim is to continue promoting the Ibn al-Haytham legacy and to extend that to many other scholars. The society includes leading academics in science, engineering, history and education from all over the world. We are very pleased to have as honorary chair the renowned Professor R Rashed. We particularly aim to organise a major international conference in 2018. This second edition will be dedicated to the successors of Ibn al-Haytham: from Kamāl al-Dīn al-Fārisī to René Descartes.

Civilisation is not only a question of acquiring knowledge and education; above all, it is a question of the ability to transform this knowledge into a productive force for economic development and the betterment of societies. This should bring together scientists and engineers from all over the world since the challenges of society are common to all humanity. Light science and technology are at the heart of these issues.

However, the true important light is that which enlightens the spirit; it is the unique renewable energy which is inexhaustible and as lasting as the existence of humanity.

Azzedine Boudrioua

Coordinator of Ibn al-Haytham International Working Group (IYL2015)
President of Ibn al-Haytham International LHiSA Society

Light : History, Science and Applications
Ibn Al Haytham International Society

Section I

History Guiding the Future, the Ibn al-Haytham Legacy

This section of the book is the main contribution and focuses on the legacy of Ibn al-Haytham.

Roshdi Rashed gives an important view of the Ibn al-Haytham research programme not only in optics but also in astronomy and statics. Ibn al-Haytham (d. after 1040) pursued the realisation of a programme to reform physical disciplines, which brought him to take up each discipline in turn. The founding action of this reform consisted in breaking from previous conception of the relationship between mathematics and physics.

Joseph Van Ess takes the example of the IYL2015 celebration to evoke light in a peaceful way. For him, Ibn al-Haytham liked to see the stars and the metropolis where he lived a millennium ago, Cairo, *al-Qāhira*, the 'Powerful' or the 'Victorious', was not yet so illuminated as are the towns of our days. The year of light is over, and enlightenment is still far away. We have been looking back to Ibn al-Haytham across one millennium, and we think that we have made some progress. However, will we have another millennium to survive?

For Michel Paty, the link maintained, throughout the history of human thought in its diversity, between *light* and *understanding*, still in use in the language, has been actually transformed, with the elaboration of scientific knowledge, in a precise meaning of each of these terms, now distinct. On the one hand, *light*, conceived as *material*, belonging to the natural world, and object of study of the science of nature specified gradually as Physics, in which Mathematics play a growing role (but without confusing their respective objects), and of which Optics is a part. On the other hand, *knowledge* and *understanding* in the proper forms of human thought, organising themselves according to the function of *reason*, which makes possible to 'transcend' the experience of the world, in different directions, including that one which makes this world *intelligible* to us, and that one also, thought to be correlative of the first, which *frees* the mind (which is given by the maintained metaphorical expression 'lights of reason').

Paul Pietquin emphasises the importance of translation. When Friedrich Risner published the very first printed edition of the Latin translation of Ibn al-Haytham's *Book of Optics* (*Kitāb al-Manāẓir*) in 1572, he chose to publish it together with another treatise, written in Europe, but largely based on Ibn al-Haytham's work: Witelo's *Perspectiva*, composed around 1270. Risner titled this collection of works *Opticae thesaurus* (Treasure of Optics), a real treasure indeed, as it remained, for more than 400 years, the only printed edition available of *Kitāb al-Manāẓir* (translated into Latin as *De aspectibus*, with the name of the author rendered as Alhacen or Alhazen). It was not until recently that critical editions of the text were published in Arabic (by Sabra) and in Latin (by Smith and by me), alongside English and French translations.

Hassan Tahiri concentrates on how to improve our understanding of the history of science and historical studies. Regarding the development of some scientific disciplines, what we sometimes know or we believe we know is not what actually happened. The modern organisation of knowledge and the structure of the teaching and research help little to narrow this gap between science as it actually produced and the dominant perception of its development. The history of astronomy, which is taken as an example, is one of the scientific disciplines most studied. Yet there is no comprehensive work on the history of its development because of the significant gaps which are still not filled. It is a project which has not yet sufficiently drawn the attention of the scientific community, because such a work requires an intercultural and collective effort.

Pierre Coulet considers the 'The problem of Ibn al-Haytham'. He begins with the simplest case that of the plane mirror. Two points A and B and a plane mirror DE being given, how to determine the point of reflection of a light ray emitted from A and reflected to B? One must then find a point C on the mirror so that the straight lines AC and BC represent the incident and the reflected rays. Later, Ibn al-Haytham provides a mechanical model of the mechanism of light reflection.

Vasudevan Lakshminarayanan presents Ibn al-Haytham's role in the field of human eye anatomy and of vision. Ibn al-Haytham was probably the first to underscore the fact that vision does not end with the formation of the image by the optics of the eye. Though Galen and others had speculated that vision involves the brain, Ibn al-Haytham was the first to explicitly state the role played by the brain. He devoted Book II of his magnum opus Kitab al-Manazir to perception. It appears as though he described or anticipated many visual and perceptual phenomena that were (re) discovered in Europe (and elsewhere) centuries later.

1 Ibn al-Haytham's Scientific Research Programme

Roshdi Rashed

For the vast majority of historians and, more generally, of laymen, Ibn al-Haytham's major contribution concerns the vision in all its aspects (physical, physiological and psychological) and especially the causes of perceptual and cognitive effects. The reform of Ibn al-Haytham, according to them, was mainly to abandon the traditional theory of vision, to a new one. Henceforth he belongs to ancient and medieval traditions, in spite of this reform, in so far that he was concerned with vision and sight.

I will argue here that this reform was a minor consequence of a more general and more fundamental research programme, and even his conception of the science of optics is quite different as so far his main task was about light, its fundamental properties and how they determine its physical behaviour, as reflection, refraction, focalisation, etc.

Some historians of optics consider that, up to the seventeenth century in Europe, the science in optics before Kepler was aimed primarily at explaining vision. The merest glance at the optical works of Ibn al-Haytham leaves no doubt that this global judgement is far from being correct. Indeed, this statement is correct as far as it concerns the history of optics before the shift done by Ibn al-Haytham and the reform he accomplished. As the successor of Ptolemy, al-Kindī and Ibn Sahl, to mention only a few, he unified the different branches of optics: optics, dioptrics, anaclastics, meteorological optics, etc. This unification was possible only for a mathematician who focused on light, and not on vision. As far as I know, nobody before Ibn al-Haytham wrote such books titled: *On Light*, *On the Light of the Moon*, *On the Light of the Stars*, *On the Shadows*, among others, in which nothing concerns sight. At the same time, three books from his famous *Book of Optics* are devoted strictly to the theory of light. None of the authors before him, who were mainly interested in vision, wrote a very important contribution on physical optics such as the one on *The Burning Sphere*.

I begin by quoting the expression which Ibn al-Haytham repeated more than once in his different writings on optics. At the beginning of this famous *Book of Optics*, he writes:

> Our subject is obscure and the way leading to knowledge of its nature difficult, moreover our inquiry requires a combination of the natural and mathematical sciences.*

---

* *The Optics of Ibn al-Haytham, Books I-III*, translated by A I Sabra, London, 1989, I, p. 4.

3

But such a combination in optics, for instance, requires one to examine the entire foundations and to invent the means and the procedures to apply mathematics on the ideas of natural phenomena. For Ibn al-Haytham, it was the only way to obtain a rigorous body of knowledge.

Why this particular turn, at that time? Let me remind you that Ibn al-Haytham lived at the turn of the first millennium. He was the heir of two centuries of scientific research and scientific translations, in mathematics, in astronomy, in statics, in optics, etc. His time was devoted to intense research in all these fields. He himself wrote more on mathematics and on astronomy than on optics *per se*. According to early biobibliographers, Ibn al-Haytham wrote 25 astronomical works: twice as many works on the subject as he did on optics. The number of his writings alone indicates the huge size of the task accomplished by him and the importance of astronomy in his life work. In all branches of mathematics, he wrote more than all of his writings on astronomy and on optics together. If he wrote on optics the famous huge book, *Kitāb al-Manāẓir (The Book of Optics)*, in astronomy, likewise, he wrote a huge book entitled *The Configuration of the Motions of Each of the Seven Wandering Stars*.

Before coming back in some details to these contributions, let me characterise Ibn al-Haytham's research programme.

1. It is a new one, concerning the relationships between mathematics and natural phenomena, never conceived before. His aim is to mathematise every empirical science. This application of mathematics can take different forms, not only given to the different disciplines, but also in one and the same discipline.
2. It does not concern only optics, but every natural science, that is, for the epoch, astronomy and statics.
3. Its success depends on the means – mathematical, linguistic and technical – by which mathematics controls the semantic and syntactical structures of natural phenomena.

Let us come now to Ibn al-Haytham's optics. As we have said above, Ibn al-Haytham was preceded by two centuries of translation into Arabic of the main Greek optical writings, as well of inventive research. Among his Arabic predecessors were al-Kindī, Qusṭā ibn Lūqā, Aḥmad ibn 'Īsā 'Uṭārid, etc. During these two centuries, the interest shown in the study of burning mirrors is an essential part of the comprehension of the development of catoptrics, anaclastics and dioptrics, as the book produced between 983 and 985 by the mathematician al-'Alā' Ibn Sahl testifies. Before this contribution of Ibn Sahl, the catoptricians like Diocles, Anthemius of Tralles, al-Kindī, etc.,* asked themselves about geometrical properties of mirrors and about the light they reflect at a given distance. Ibn Sahl modifies the question by consider-

---

* See R Rashed, *Les Catoptriciens grecs.* I: *Les miroirs ardents*, édition, traduction et commentaire, Collection des Universités de France, Paris: Les Belles Lettres, 2000; and R Rashed, *Œuvres philosophiques et scientifiques d'al-Kindī.* Vol. I: *L'Optique et la Catoptrique d'al-Kindī*, Leiden: E J Brill, 1997; Arabic translation: *'Ilm al-manāẓir wa-'ilm in'ikās al-ḍaw'*, Silsilat Tārīkh al-'ulūm 'inda al-'Arab 6, Beirut: Markaz Dirāsat al-Waḥda al-'Arabiyya, 2003.

ing not only mirrors but burning instruments, that is, those which are susceptible to light not only by reflection, but also by refraction; and how in each case the focalisation of light is obtained. Ibn Sahl then studies, according to the distance of the source (finite or infinite) and the type of lighting (reflection or refraction), the parabolic mirror, the ellipsoidal mirror, the plano-convex lens and the biconvex lens. In each of these, he proceeds to a mathematical study of the curve, and then expounds a mechanical continuous drawing of it. For the plano-convex lens, for instance, he starts by studying the hyperbola as a conic section, in order to take up again a study of the tangent plane to the surface engendered by the rotation of the arc of hyperbola around a fixed straight line, and, finally, the curve as an anaclastic curve, and the laws of refraction.

These studies focused on light and its physical behaviour were instrumental in the discovery by Ibn Sahl of the concept of a constant ratio, characteristic of the medium, which is a masterpiece in his study of refraction in lenses, as well as his discovery of the so-called Snellius' law.

Thus, Ibn Sahl had conceived and put together an area of research into burning instruments and also anaclastics. However, obliged to think about conical figures other than the parabola and the ellipse – the hyperbola, for example – as anaclastic curves, he was quite naturally led to the discovery of the law of Snellius.

Rich in technical material, this new discipline is in fact very poor on physical content: it is faint and reduces a few energy considerations. By way of example, at least in his writings that have reached us, Ibn Sahl never tried to explain why certain rays change direction and are focused when they change medium: it is enough for him to know that a beam of rays parallel to the axis of a plano-convex hyperbolic lens gives by refraction a converging beam. As for the question why the focusing produces a blaze, Ibn Sahl is satisfied with a definition of the luminous ray by its action of setting ablaze by postulating, as did his successors elsewhere for much longer, that the heating is proportional to the number of rays.

While Ibn Sahl was finishing his treatise on *Burning Instruments* very probably in Baghdad, Ibn al-Haytham was probably beginning his scientific career.

Compared with the writings of the Greek and Arab mathematicians who preceded him, the optical work by Ibn al-Haytham presents at first glance two striking features: extension and reform. It will be concluded on a more careful examination that the first trait is the material trace of the second. In fact, no one before Ibn al-Haytham had embraced so many domains in his research, collecting together fairly independent traditions: mathematical, philosophical, medical. The titles of his books serve moreover to illustrate this large spectrum: *The Light of the Moon*, *The Light of the Stars*, *The Rainbow and the Halo*, *Spherical Burning Mirrors*, *Parabolical Burning Mirrors*, *The Burning Sphere*, *The Shape of the Eclipse*, *The Formation of Shadows*, *On Light*, us well as his *Book of Optics* translated into Latin in the twelfth century and studied and commented on in Arabic and Latin until the seventeenth century. Ibn al-Haytham therefore started not only the traditional themes of optical research but also other new ones to cover finally the following areas: optics, catoptrics, dioptrics, physical optics, meteorological optics, burning mirrors, the burning sphere.

A more careful look reveals that, in the majority of these writings, Ibn al-Haytham pursued the realisation of his programme to reform the discipline, which brought clearly to take up each different problem in turn. The founding action of this reform consisted in making clear the distinction, for the first time in the history of optics, between the conditions of propagation of light and the conditions of vision of objects. It led, on the one hand, to providing physical support for the rules of propagation – it concerns a mathematically guaranteed analogy between a mechanical model of the movement of a solid ball thrown against an obstacle, and that of the light – and, on the other hand, to proceeding everywhere geometrically and by observation and experimentation. It led also to the definition of the concept of light ray and light bundle as a set of straight lines on which light propagates, rays independent from each other which propagate in a homogeneous region of space. These rays are not modified by other rays which propagate in the same region. Thanks to the concept of light bundle, Ibn al-Haytham was able to study the propagation and diffusion of light mathematically and experimentally. Optics no longer has the meaning that is assumed formerly: a geometry of perception. It includes henceforth two parts: a theory of vision, with which are also associated a physiology of the eye and a psychology of perception, and a theory of light, to which are linked geometrical optics and physical optics. The organisation of his *Book of Optics* reflects already the new situation. In it are books devoted in full to propagation – the third chapter of the first book and Books IV to VII; others deal with vision and related problems. This reform led to, among other things, the emergence of new problems, never previously posed, such as the famous 'problem of Alhazen' on catoptrics, the examination of the spherical lens and the spherical dioptre, not only as burning instruments but as optical instruments, in dioptrics; and to experimental control as a practice of investigation as well as the norm for proofs in optics and more generally in physics.

Let us follow now the realisation of his reform in the *Book of Optics* and in other treatises. This book opens with a rejection and a reformulation. Ibn al-Haytham rejects straightaway all the variants of the doctrine on the visual ray, to ally himself with philosophers who defended a doctrine of intromission on the form of visible objects. A fundamental difference remains nevertheless between him and the philosophers, such as his contemporary Avicenna: Ibn al-Haytham did not consider the forms perceived by the eyes as 'totalities' which radiate from the visible object under the effect of light, but as reducible to their elements; from every point of the visible object radiate a ray towards the eye. The latter has become without soul, without πνεῦμα ὀπτικόν, a simple optical instrument. The whole problem was then to explain how the eye perceives the visible object with the aid of these rays emitted from every visible point.

After a short introductory chapter, Ibn al-Haytham devotes two successive chapters – the second and the third books of his *Book of Optics* – to the foundations of the new structure. In one, he defines the conditions for the possibility of vision, while the other is about the conditions for the possibility of light and its propagation. These conditions, which Ibn al-Haytham presents in the two cases of empirical notions, that is, as resulting from an ordered observation or a controlled experiment, are effectively constraints on the elaboration of the theory of vision, and in this way on the new style of optics. The conditions for vision detailed by Ibn al-Haytham are six: the visible object must be luminous by itself or illuminated by another; it must be opposite the eye, that is, one can draw a straight line to the eye from each of its points; the medium that separates it from the eye must be transparent, without being cut into by any opaque obstacle; the visible object must be more opaque than this medium; it must be of a certain volume, in relation to the visual sharpness. These are the notions, writes Ibn al-Haytham, 'without which vision cannot take place'. These conditions, one cannot fail to notice, do not refer, as in the ancient optics, to those of light or its propagation. Of these, the most important, established by Ibn al-Haytham, are the following: light exists independently of vision and exterior to it; it moves with great speed and not instantaneously; it loses intensity as it moves away from the source; the light from a luminous source – substantial – and that from an illuminated object – second or accidental – propagate onto bodies which surround them, penetrate transparent media, and light up opaque bodies which in turn emit light; the light propagates from every point of the luminous or illuminated object in straight lines in transparent media and in all directions; these virtual straight lines along which light propagates form with it 'the rays'; these lines can be parallel or cross one another, but the light does not mix in either case; the reflected or refracted light propagates along straight lines in particular directions. As can be noted, none of these notions relate to vision.

A theory of vision must henceforth answer not only the six conditions of vision, but also the conditions of light and its propagation. Ibn al-Haytham devotes the rest of the first book of his *Book of Optics* and the two following books to the elaboration of this theory, where he takes up again the physiology of the eye and a psychology of perception as an integral part of this new theory of intromission.

Three books of the *Book of Optics* – the fourth to the sixth – deal with catoptrics.

Let us consider some aspect of this research into catoptrics by Ibn al-Haytham. He restates the law of reflection, and explains it with the help of the mechanical model already mentioned. Then he studies this law for different mirrors: plane, spherical, cylindrical and conical. In each case, he applies himself above all to the determination of the tangent plane to the surface of the mirror at the point of incidence, in order to determine the plane perpendicular to this last plane, which includes the incident ray, the reflected ray and the normal to the point of incidence. Here, as in his other studies, to prove these results experimentally, he conceives of and builds an apparatus inspired by the one that Ptolemy constructed to study reflection, but more complicated and adaptable to every case. Ibn al-Haytham also studies the image of an object and its position in the different mirrors. He applies himself to a whole class of problems: the determination of the incidence of a given reflection in the different mirrors and conversely. He also poses for the different mirrors the problem with which his name is associated: given any two points in front of a mirror, how does one determine on the surface of the mirror a point such that the straight line which joins the point to one of the two given points is the incident ray, while the straight line that joins this point to the other given point is the reflected ray. This problem, which rapidly becomes more complicated, has been solved by Ibn al-Haytham.

Ibn al-Haytham pursues this catoptric research in other essays, some of which are later than the *Book of Optics*, such as *Spherical Burning Mirrors*.* It is in this

---

* *Fī al-marāyā al-muḥriqa bi-al-dawā'ir*, ms. Berlin, Oct. 2970/7, fols 66ʳ–73ᵛ.

essay of a particular interest that Ibn al-Haytham discovers the longitudinal spherical aberration.

Ibn al-Haytham devotes a substantial part of the seventh book to the study of the image of an object by refraction, notably if the surface of separation of the two media is either plane or spherical. It is in the course of this study that he settles on the spherical dioptre and the spherical lens, following thus in some way the research by Ibn Sahl, but modifying it considerably; this study of the dioptre and the lens appears in effect in the chapter devoted to the problem of the image, and is not separated from the problem of vision. For the dioptre, Ibn al-Haytham considers two cases, depending on whether the source – punctual and at a finite distance – is found on the concave or convex side of the spherical surface of the dioptre.

Ibn al-Haytham studies the spherical lens, giving particular attention to the image that it gives of an object. He restricts himself nevertheless to the examination of only one case, when the eye and the object are on the same diameter. Put another way, he studies the image through a spherical lens of an object placed in a particular position on the diameter passing through the eye. His procedure is not without similarities to that of Ibn Sahl when he studied the biconvex hyperbolic lens. Ibn al-Haytham considers two dioptres separately, and applies the results obtained previously. It is in the course of his study of the spherical lens that Ibn al-Haytham returns to the spherical aberration of a point at a finite distance in the case of the dioptre, in order to study the image of a segment which is a portion of the segment defined by the spherical aberration.

Let us stop at this point on spherical aberration to conclude.

With Ibn al-Haytham, one result has been definitively obtained: the half century which separates him from Ibn Sahl should be counted among the distinctive moments in the history of optics: dioptrics appears to have extended its domain of validity and, by its very progress, to have changed its orientation. With Ibn al-Haytham, the conception of dioptrics as a geometry of lenses has become outdated. Here again, in his own words, we must combine mathematics and physics in order to study dioptres and lenses, whether burning or not. The mathematisation could only be achieved with Ibn al-Haytham because he separated the study of the natural phenomenon of light from that of vision and sight. In optics as in astronomy, the research programme of Ibn al-Haytham is the same: mathematise the discipline and combine this mathematisation with the ideas of the natural phenomena.

2 Light and Vision before Ibn al-Haytham
The Perspective of Islamic Theology

J Van Ess

Light is a benefit for mankind; this is why the nations of this world – or their diplomats – decided to celebrate it for an entire year. Light was the first thing to be created by God, as we are told in the *Book of Genesis*, and we all remember how Haydn in his oratorio *Die Schöpfung* prepares us for the solemn moment when, after a mysterious pianissimo, the full orchestra and the entire choir rejoice after having reached this word of the text, 'Licht', C major and fortissimo. God was still alone then and Light the first demonstration of His power. The universe did not yet exist; only two days later, according to the Biblical report, the sun and the moon were created. We have to imagine that those who rejoice in the event are angels, not human beings. When, however, in their stead man, 'le maître de la nature' as he was called by Descartes, made an attempt at demonstrating his power in the same way, recently, only 70 years ago, there was not so much to rejoice in. I still remember when, in 1945 after the war had come to an end in Europe, a light 'brighter than a thousand suns' ('heller als tausend Sonnen' to quote the title of Robert Jungk's famous book of 1956) was produced in East Asia, above the town of Hiroshima,* light as *force de frappe* to use the French expression, surpassing in its efficiency everything we had experienced before, with thousands of victims. Afterwards, the crime could only be justified as a test; in contrast to God, man can never be sure of his exploits right away. As a matter of fact, there was a second 'test' shortly afterwards (Nagasaki), and there would have been a third one, on Tokyo, if the Tenno had not been so wise to give in. When man is left alone with his creativity, he is incalculable ('quand il veut faire l'ange il fait la bête', as said Pascal†), and only researchers of the forgotten past applied their tests to themselves (Madame Curie, for instance). When we produce too much light, the benefit cannot be taken for granted, and certainly we don't see the stars any more.

---

* Jungk (1913–1994), a German Jew born in Berlin, wrote this book in German: *Heller als tausend Sonnen.* He had returned to Germany in 1945 after having lived in exile first in England and then in Switzerland. The English translation appeared 1958 in New York: *Brighter than a Thousand Suns. A Personal History of the Atomic Scientists.* The title of the book is a quotation from a religious text, the *Bhagavadgita.*
† *Pensées* III 2, nr. 358.

The man whose memory we are going to evoke for a day or two dealt with light in a peaceful way. Ibn al-Haytham (died shortly after 430/1039) liked to see the stars, and the metropolis where he lived one millennium ago, Cairo, *al-Qāhira*, the 'Powerful' or the 'Victorious', was not yet so shiny as are the towns of our days.[*] He was good in mathematics and spherical trigonometry, skills useful not only for astronomy but also for religious purposes, the timekeeping of the daily prayer and the computation of its direction towards Mecca.[†] A colleague of his, a man by the name of Ibn Yūnus who was older than he (he died in 1009)[‡] and whom he may have no longer met, had established trigonometrical tables which were to last for centuries and were named after the caliph in power at that moment, al-Ḥākim[§] (a man who became the hero of popular stories[¶] still recorded by Gérard de Nerval when he travelled to Egypt in 1843).[**] Ibn al-Haytham, Alhacen in the Latin tradition, produced an astronomical handbook[††] which also continued to be used for centuries and ended up by being translated into Latin, Hebrew, Castilian and Persian. However, it seems to be an early work; he does not yet mention the great celestial event of his time which ran counter to the rules of the prevailing system: the explosion of a supernova in the year 1006.[‡‡] His originality did not appear in astronomy anyway; it came forward in another discipline: optics.[§§]

---

[*] Cairo was founded in 359/970 when the Fātimid caliph al-Mu'izz (reigned 341/953–365/975) had conquered Egypt. The town was established as an annex to the old Greco-Coptic settlement Fusṭāṭ (=φοσσάτον < lt. *fossatum* 'encampment surrounded by a ditch') which is nowadays called 'Old Cairo'. The name *al-Qāhira* possibly alludes to al-Mu'izz himself who in an encomium is called *al-qahhār*, 'the conqueror' or 'the victor'. Originally the settlement had been called Manṣūriyya. Later generations who lived when the Fātimids had been replaced by the Ayyubids (and after them by the Mamluks) preferred explaining the name by a fanciful story according to which the astrologers who had to find an auspicious date for the official founding ceremony chose a day when the planet Mars (=*al-qāhir* 'the Powerful') was the ascendent (*ṭāli'*; cf. for instance Maqrīzī, *Le Manuscrit autographe d'al-Mawā'iẓ wal-i'tibār*, ed. Ayman Fu'ād Sayyid, London 1416/1995, p. 38, 6 f.). But this is a myth; Mars was not called *al-qāhir* in astrology (cf. P Kunitzsch, *Zur Namengebung Kairos*, in: Der Islam 52/1975/209 ff.).

[†] Cf. D A King in EI² VII 27 ff. s. v. *Mīḳāt* and V 83 ff. s. v. *Ḳibla*.

[‡] Cf. B R Goldstein in EI² III 969 f. and D A King in *Dictionary of Scientific Biography* XIV 574 ff. s. n. *Ibn Yūnus*.

[§] The *Zīdj al-Ḥākimī al-kabīr*; cf. D A King and J Samsó in EI² XI 499b s. v. *Zīdj*. Ibn Yunus had an observatory at his disposal, but the account according to which this institution had been equipped for him by the caliph himself is legendary (cf. J Samsó in EI² VI 600b s. v. *Marṣad*). Al-Ḥākim (born in 375/985) was still young at that time, and until 390/1000 he had been under the tutelage of a eunuch who had managed to push aside the Berber entourage which had dominated the palace before; cf. EI² I 1041 f. s. n. *Bardjawān* (B Lewis).

[¶] He may have been the prototype for the stories in the *Tales of the 1001 Nights* where the Abbasid caliph Hārūn al-Rashīd appears as roaming through the streets of Baghdād at night.

[**] *Voyage en Orient* (published in 1851), vol. II, part 3: *Histoire du Calif Hakem*.

[††] The *Kitāb fī Hay'at al-'ālam*, analyzed by M Schramm, *Ibn al-Haitams Weg zur Physik* (Wiesbaden 1963), p. 63 ff.

[‡‡] The 'forelock star' (*kawkab al-dhu'āba*) of our sources; for further detail cf. J van Ess, *Chiliastische Erwartungen und die Versuchung der Göttlichkeit. Der Kalif al-Ḥākim (386–411 H.)*; Heidelberg 1977, p. 40 ff. For the early date of the *Kitāb fī Hay'at al-'ālam* cf. also A Sabra in: *Dictionary of Scientific Biography* VI 197b; the book was not the only manual composed by Ibn al-Haytham, but perhaps the first one.

[§§] Ibn Khaldūn called him the most famous representative of this discipline in the Islamic world (*Muqaddima*, trs. Abdesselam Cheddadi, *Le Livre des Exemples* I, Paris 2002, p. 956).

He may have settled in Cairo rather late. He was not an Egyptian by birth. His parents had lived in or around Baṣra, the port at the Persian Gulf, and he himself may have been of Persian origin.* He spent some time in Ahwāz, a mountanous area on the way to the Iranian tableland – for us, the scene of Saddam Husayn's first Gulf War when he still enjoyed the sympathies of the 'West'. Islam had created a world of high mobility, not only due to the institution of the Ḥajj but also because of vivid scholarly communication. Ibn al-Haytham may have come on invitation; he arrived with a big project in mind. However, he failed. He had intended or proposed to regulate the flow of the Nile. He knew the problem from Iraq where the lowlands around the Euphrates and the Tigris used to be afflicted by horrible inundations which caused the rivers to change their beds and constantly enlarged the alluvial plain in the Shaṭṭ al-'Arab beyond the port of Baṣra. However, the situation in Egypt was different. In Iraq, the flood originated in the mountains of Upper Mesopotamia, and the water came down irregularly, by surprise; Noah had built his ark there in order to escape the Deluge. In contrast to this, the Nile regularly increased by swelling every year; the reason for this, the monsoon rainfall in what is now Ethiopia, was not known to everybody and *a fortiori* not to somebody who came from abroad. Ibn al-Haytham is supposed to have travelled up to Aswān, that is, the first cataract, until he recognised that his project could not be realised. Our sources pretend that when he arrived from Baṣra, the caliph had come to bid him welcome; he seems to have already had a certain reputation, and the caliph is even said to have sent him money to Baṣra in order to cover his travel expenses.†

Al-Ḥākim may have been disappointed. In any case, he was incalculable – like any ruler. So why did Ibn al-Haytham stay?‡ Possibly because of the excellent working conditions. For some decades, Cairo had become the hub of the Islamic world,§ and in 1005 a 'House of Wisdom' was opened there, a kind of research centre where scholars of all disciplines could come to read and work, theologians and jurists, of course, but also physicians, astronomers, mathematicians and so forth. They were offered ink and paper, and there was a staff, librarians, etc., who could give a helping

---

* The population of Baṣra was very mixed, and we know that a considerable number of people belonging to the lower classes were native speakers of Persian, in a variety which was just acquiring its written modern form. But we do not have any information about Ibn al-Haytham's family.

† Strictly speaking we find this report only in one source which dates from a period two centuries after Ibn al-Haytham's death: Ibn al-Qifṭī's (568/1172–646/1248) *Ta'rīkh al-ḥukamā'* (ed. J Lippert, Leipzig 1903), p. 166. The full title of the book is *Ikhbār al-'ulamā' bi-akhbār al-ḥukamā'*, but it survived only in an epitome produced in 647/1249 (i.e., shortly before the conquest of Baghdad by the Mongols). The story may be legendary in some of its details, but it was then reproduced in Ibn Abī Uṣaybi'a's *'Uyūn al-anbā' fī ṭabaqāt al-aṭibbā'* which was completed in 667/1269 (ed. August Müller, Königsberg 1884, vol. II 90, -7 ff.). Another author who mentions the event is Ẓahīraddīn al-Bayhaqī who lived more than half a century before Ibn al-Qifṭī (ca. 493/1097–565/1169–1170), in his *Tatimmat Ṣiwān al-ḥikma* (ed. M Kurd 'Alī, Damascus 1365/1946, ²1396/1976, under the title of *Ta'rīkh ḥukamā' al-Islām*), pp. 85–88 nr. 39. But Bayhaqī lived in Iran, mostly in the town of Nīshāpūr, and he was not well informed. He is very brief, and he seems to have never been in Egypt.

‡ If he stayed at all. Bayhaqī pretends that he left for Syria and that a nobleman from Simnān in Persia came to study with him there (p. 86, 6 ff.). But this story has mainly the function of emphasising Ibn al-Haytham's ascetic lifestyle and his contempt of money, for he is said to have wanted 100 Dinar per month for his teaching but then to have returned the money when his student had finished the course and left again. There is no reason for denying that Ibn al-Haytham lived for a long time in Cairo.

§ Cf. Heinz Halm, *Die Kalifen von Kairo. Die Fāṭimiden in Ägypten 973–1074* (Munich 2003).

hand.* This was certainly attractive for foreigners†; Ibn al-Haytham did not only want to study, he also wanted to make experiments, and he needed somebody who would help him construct a camera obscura.‡ The libraries of the palace were very rich; when, 60 years later, they were looted because the state had gone bankrupt and could not pay its soldiers any more, an incredible amount of volumes got lost, 18,000 alone in the Greek sciences according to one of our sources.§ The figure is probably too high. Just imagine: the library of the monastery of Benediktbeuren in Germany contained only slightly more than 100 volumes in the year 1032, and most of it was religious stuff.¶ Not only was the Islamic world richer and more splendid than Europe, its civilisation was also completely secular, not clerical and certainly not 'fundamentalist'. Fundamentalism is a modern phenomenon, in Islam as well as in the 'West'.** It is true that everybody who lived under the umbrella of Islam at that period was, in a way, religious, but what he really set his heart on could easily be something different: poetry, literature – or scientific material inherited from Late Antiquity.

We do not know to what extent Ibn al-Haytham could profit from this overwhelming supply. He is said to have earned his living by copying scientific texts and selling them, books which were difficult to understand and required high precision because they contained calculations and diagrams.†† We still have one of these manuscripts from his hand; it contains the Arabic translation of Apollonius's *Conics*, and he copied it in April 1024, that is, after al-Ḥākim's death.‡‡ The market he worked for was

---

* Halm 206 ff.; also van Ess, *Chiliastische Erwartungen* 33 f.

† Nāṣir-i Khosraw came from distant Transoxania to see the town (and his Fāṭimid *imām*); he arrived in 439/1047, that is, some years after Ibn al-Haytham's death.

‡ For the camera obscura cf. Schramm, *Ibn al-Haytam's Weg zur Physik* 200 ff.; but Ibn al-Haytham also needed implements for his observations of the moon (ib. 146 ff.).

§ Halm 207 and 412. The Greek texts were found there in Arabic translation, of course, and in company with items of the enormous secondary literature produced by the Arabs themselves. In addition to these manuscripts 2,400 magnificent reproductions of the Qur'ān are said to have been stolen on this occasion. The stuff was probably sold on the market, but book bindings could sometimes also be used by uneducated people as soles for their sandals (Halm 412, according to Maqrīzī).

¶ This comparison was made by Adam Mez, in his *Die Renaissance des Islams* (Heidelberg 1922), p. 165.

** As is well known the term 'fundamentalism' came up in the United States where, in Christian Protestantism, the 'fundamentals' were originally understood in a positive sense. For the topic cf. now Michael Cook, *Ancient religions, modern politics* (Princeton UP 2014); the author takes his examples from the Islamic world, from India, and from South American Catholicism, but he omits North America.

†† A Jewish contemporary of Ibn al-Qifṭī who lived in Aleppo ('Yūsuf al-Fāsī al-Isrā'īlī' = Josef b. Yehuda Ibn Aknīn, d. 1226, a disciple of Maimonides) pretended to have heard that Ibn al-Haytham copied every year Euclid's *Elements* and Ptolemy's *Almagest*, together with the texts which were read 'in-between' in the Greek syllabus (*al-mutawassiṭāt*, for instance the *Sphaerica* by Menelaus), and that he could sell these books for the sum of 150 Egyptian dīnārs (Ibn al-Qifṭī, *Ta'rīkh al-ḥukamā'* 167, 8 ff. > Ibn Abī Uṣaybi'a II 91, 17 ff.).

‡‡ Cf. the facsimile of the colophon in: Schramm, *Ibn al-Haytams Weg*, p. xi. Ibn al-Qifṭī reports to have seen another autograph written in 432/1041 (*Ta'rīkh al-ḥukamā'* 167, 14 f., an important testimony for fixing Ibn al-Haytham's death date). The physician Ibn Riḍwān (388/998–453/1061), younger than Ibn al-Haytham by one generation (cf. EI² III 906 f. s. n.), tells us that he copied Ibn al-Haytham's treatise on the light of the moon in mid-Sha'bān 422/first half of August 1031 (Ibn al-Qifṭī, ib. 444, 14 ff.), that is, at a moment when Ibn al-Haytham was still alive and he himself perhaps still getting his living as an astrologer on the streets of Cairo (cf. van Ess, *Chiliastische Erwartungen* 42).

relatively small, and part of his production may have gone to the 'House of Wisdom'. He could expect to get a good price, but he never became a rich man. Compared to the indigenous religious elite who emerged from the old families, he was isolated, a foreigner without any inherited connections.* However, Islamic society was rather permeable, a society without an established nobility, and he was a social climber. Ibn al-Haytham had this in common with other great figures of his time: Avicenna, for instance, who died a few years before him,[†] or Bīrūnī, the great scientist and polymath who learnt Sanskrit and wrote a famous book about India, the first description of a foreign civilisation in the history of mankind we know of. Bīrūnī even seems to owe his name to his low origin; he grew up in the outskirts of his native town (Kāth, the capital of the Afrīghid Khwārazmshāhs), in the slums or bidonvilles 'outside' (*bērūn* in Persian).[‡] Ibn al-Haytham may have never heard about these figures; they lived far away in the East, in Central Asia, Avicenna in Bukhārā and Bīrūnī in Khoresmia, south-east of the Caspian Sea.[§] However, in our context, they are worth mentioning because in the year 999, when both of them were still very young, they got engaged in a correspondence which is still preserved and tells us something about the spirit of the time. It was the older one of the two who asked the questions: Bīrūnī, at the age of 28, and Avicenna, 18 years old, was already mature (and arrogant) enough to furnish the answers. The questions are bold and mark the horizon of contemporary scholarly endeavour: Are there other solar systems out there among the stars, or are we alone in the universe? Or: Has the earth been created whole and complete, or has it evolved over time? Or: Is all motion of celestial bodies linear, as Aristotle maintained, or is it elliptical and would motion in this form be possible?[¶] This was not necessarily the kind of questions which were asked in Fāṭimid Cairo, but the point is that nobody seems to have been vituperated or ostracised when he asked them, and nor would any institution have intervened.[**] Neither of the two partners had to fear for his life; on the contrary, they both made a remarkable career at Iranian courts.

Like Bīrūnī, Ibn al-Haytham was mainly an empiricist. True knowledge only derives from what we perceive by our senses, he said.[††] The statement was inspired

---

* After al-Ḥākim's death he lived in a chamber under the dome of one of the gates of the al-Azhar mosque, that is, in the neighbourhood of the Fāṭimid palace area (Ibn al-Qifṭī 167, 5 f.). He had been married, for we know about a son in law of his (see n. ‡, p. 21).

[†] In 428/1037; cf. EI² III 941 ff. s. v. *Ibn Sīnā* (A-M Goichon) and Encyclopaedia Iranica III 66–110 (several authors).

[‡] He died in 440/1048; cf. EI² I 1236 ff. (D J Boilot), EIr IV 274–287 (several authors), and DSB II 147–158 (E S Kennedy). A short summary, together with a discussion of the exact death date, also in: van Ess, *Der Eine und das Andere* (1–2, Berlin 2011), vol. I, p. 725 ff.

[§] Bīrūnī owed his career to the patronage of a Khwārazmian aristocrat, Abū Naṣr Ibn ʿIrāq who was an astronomer and mathematician himself (cf. B R Goldstein in EI² III 808 s. n. *Ibn ʿIrāq* and D Pingree in EIr I 351 f. s. n. *Abū Naṣr Manṣūr*).

[¶] The correspondence has been edited by S H Naṣr and M Mohaqqeq: *Abū Rayḥān Bīrūnī va-Ibn Sīnā, al-asʾila wal-ajwiba* (Tehran 1352 š./1973). Cf. now S F Starr, *Lost Enlightenment. Central Asia's Golden Age from the Arab Conquest to Tamerlane* (Princeton UP 2013), p. 1 ff.

[**] Starr compares this to the fate of Giordano Bruno (1548–1600) who, 'six hundred years later, was burned at the stake for championing the plurality of worlds' (p. 1).

[††] This sentence was already quoted by Moritz Steinschneider, in: Bolletino di bibliografia e di storia delle scienze matematiche e fisiche 5/1872, p. 466. It is part of an autobiographical text which was first translated (into German) by Eilhard Wiedemann, *Ibn al-Haiṯam, ein arabischer Gelehrter* (in: Festschrift J Rosenthal, Leipzig 1906, pp. 149–178).

by Galen whom he quotes in the same context. However, he believed in what he said. He did not feel at home with dogmatic positions as they could be found in religious discourse, and he had read an enormous amount of Greek texts which he proudly enumerates in his bibliography.* However, I do not want to repeat what has been said before.† I would rather like to concentrate on one question: Why was it just the concept and nature of light where Ibn al-Haytham, in all his stunning erudition, started on a new departure? Was there something in his Islamic 'cultural memory' which made this topic more attractive than other ones?

The Quran does not contain a Genesis report; so there is no talk about the creation of light. However, in sura 24:35, God is said to be 'the Light of Heaven and Earth', and the words which follow, a comparison with a lamp in a niche, gave the exegetes a lot to think about. The passage left no doubt that light was something beneficial: 'Light upon Light. God guides to his Light whom He will'. Light, in its abundance, offers orientation; it marks the right direction. It is not primarily a demonstration of God's power. This latter aspect is made explicit in other passages, and in a different way. Take sura 7:143, for instance, where Moses wants to see God on Mount Sinai and is told: 'Thou shalt not see Me. But behold the mountain – if it stays fast in its place, then thou shalt see Me' (sura 7:143). This was an *argumentum e contrario*, for God, as the text continues, made the mountain 'crumble to dust, and Moses fell down swooning'. The text does not say that it was God's Light (a lightning perhaps) which made Mount Sinai crumble, but we are made to believe just this since Moses had been warned to demand something he could see, that is, a vision. Everybody on the Arabian Peninsula knew that it was very painful to look into the light of the sun. Therefore, theophany was at this moment rather achieved by way of the word, that is, the Law. This is why Moses had survived the experience.

---

* He takes Aristotle's universality as an example and wants to show that, being 63 lunar years old at that moment, he had succeeded in emulating this ideal. This is why he included the bibliography in the autobiographical *Risāla* mentioned above. Ibn al-Haytham enumerates everything which he wrote up to the end of the year 417 H., 'waiting for old age to come' (*'uddatan li-zamān al-shaykhūkha wa-awān al-haram*): 70 titles altogether plus those publications which got lost or which he could not complete for certain reasons. The paradigmatic character of the treatise is evident, with 70 as a symbolic figure. It is usually quoted on the authority of Ibn Abī Uṣaybi'a (*'Uyūn al-anbā'* II 91, 22 ff.), but we have now a separate version where the introduction and the colophon are preserved and which seems to have been the 'Vorlage' used by Ibn Abī Uṣaybi'a; it was copied from the original (the autograph?) in the year 556/1161 and thus preceded the *'Uyūn al-anbā'* by more than one century. This text survived in a private collection at Lahore and was edited by A Heinen in: Festschrift H R Roemer, Beirut 1979, pp. 254–277. In both places the *Risāla* contains a second list of books which covers the years 417–419 H., a period when Ibn al-Haytham had possibly left Egypt for Iraq again. He obviously took Galen's Πίναξ as an example which had been translated into Arabic (*De libris propriis* = Περὶ τῶν ἰδίων βιβλίων γραφή; Kühn XIX 2, pp. 8–48; cf. M Ullmann, *Die Medizin im Islam*, Leiden 1970, p. 35 nr. 1). But when he talks about Galen and his enormous productivity (Heinen 267, 8 ff. > IAB II 95, -4 ff.) he refers to one of Galen's major works, his *K Ḥīlat al-bur'*, that is, his *Megatechne/Ars magna = K aṣ-Ṣinā'a al-kabīra* (Kühn X 1, pp. 1–1021; Ullmann 45 nr. 39). It is there that his sceptical statement about the dogmatic positions (*i'tiqādāt*) is found (Heinen 259, 4 ff. > Ibn Abī Uṣaybi'a II 92, 3 ff.). Ibn Abī Uṣaybi'a may have included him in his book (which is, after all, about physicians, *aṭibbā'*) just for this reason; Ibn al-Haytham seems to have never practiced as a medical doctor.

† For further detail cf. the magnificent article by Abdalhamid Sabra in: DSB VI 189 ff.; also R Rashed in: H Selin (ed.), *Encyclopedia of the History of Science* (Berlin, Springer 2008), vol. I, pp. 1090–1093 s. v. *Ibn Al-Haytham* and E Kheirandish, ib. II 1793–1796 s. v. *Optics in the Islamic World*.

God obviously did not want him to lose his life by being exposed to His light; He rather spoke to him.

However, why does God want then to be understood as the 'Light' of Heaven and Earth and not only as their Lord? We have, of course, to take historical tradition into account, Jewish as well as Christian. In the Apostles' Creed, Christ, the second person of the Trinity, is described as *lumen de lumine* 'light derived from (God's) light'. It all started, however, in Iran; light as an active principle in cosmology as well as in everyday life was an ancient Zoroastrian idea. The difference was that it always showed up there in a mixture with a second principle which competed with it: Darkness, Ahriman, the Devil. This dualistic model had the advantage that one did not have to explain evil; evil is simply there, every day, and we only have to become aware of it. The devil is a reality which helps us to understand our world more easily; we do not have to excuse ourselves all the time. Only when dualism is replaced by monotheism does one get stuck with the question of theodicy: Why does God the Almighty allow evil to happen? You may remember that Voltaire, in his ironical way, dealt with this conundrum in his novel *Candide ou l'optimisme*, after the earthquake of Lisbon in the year 1755. In our days, this way of subdividing the realities of our life into good ones and bad ones is sometimes disposed of as 'manichean'. However, we do not have to look so far beyond our own borders. The best examples are found in American fundamentalism, at least when it comes to politics; there our ears have got accustomed to expressions like the 'axis of Evil' or the 'Evil Empire'. When we then fight against such a phenomenon, we do so 'for a good cause', and the end justifies all means (even drones).

Early Islam reacted differently. We have a few fragments of a strange text which was composed in the first half of the second century H., around 750, probably in Iraq and only a little while after the Abbasid revolution. It is attributed to Ibn al-Muqaffaʿ, an Iranian intellectual who played an important part in early Arabic literature; he translated Sasanid texts – or had them translated in his office – from Middle Persian into Arabic: the Sasanid Book of Kings, moral treatises and, above all, the fables of the two jackals Kalīla and Dimna which had ultimately been derived from the Indian *Pancatantra.* He was a convert; his father had still been the adherent of an Iranian religion and had worked as a tax collector for the new Arab overlords. The text starts with the invocation: 'In the name of Light, the Merciful, the Compassionate' (*Bismi l-nūri l-raḥmāni l-raḥīm*), a parody of *bismi llāhi l-Raḥmāni l-raḥīm* – or rather: an attempt at creating a theology of light where Allāh is replaced by Light. Now we wait for Darkness, the 'evil empire', to appear. However, this does not happen; the text continues with the words: 'Praised and holy be the Light! Whoever does not know Light will not know anything else, and who has doubts about it will not have any certitude afterwards'. On the epistemological level, light is thus supposed to have an enlightening effect. We should, of course, never forget that we have nothing but a few fragments, and I won't discuss here the difficult question of authenticity.* Suffice it

---

* The fragments are only preserved in the context of a refutation, by the early Zaydi theologian al-Qāsim b. Ibrāhīm al-Rassī (ed. M Guidi, *La lotta tra l'Islam e il Manicheismo.* Rome 1927). They are translated in: van Ess, *Theologie und Gesellschaft im 2. und 3. Jahrhundert Hidschra* (1–6, Berlin 1991–1997), vol V, pp. 104–108; a commentary is found ib., vol. II, p. 29 ff.

to say that we seem to be dealing with a basically monotheist adaptation of a dualist model. God as Light was one possibility of conceptualising the ineffable reality of the Divine. However, here this is done with reference to God's mercy (*al-Raḥman al-raḥīm*), not so much with regard to His power as in the case of Moses.

Moses was the past anyway. He had not been the last prophet. The Muslims felt respect for him, but they could think of a closer relationship to God than his one. When they compared him to Muḥammad, he was simply a forerunner, and the vision of God which had been forbidden to him became a possible alternative with respect to their own prophet. In sura 17 (where there is some talk about him), the very beginning (i.e., verse 1) was supposed to deal with Muḥammad instead who had undertaken a heavenly journey at the end of which he had encountered God. This experience exalted him over all his predecessors. Two of them are afterwards explicitly mentioned in this context*: Abraham since he was called the 'friend of God' (*khalīl Allāh*), and Moses since he had been talked to by God (*kalīm Allāh*). It was in contrast to them that Muḥammad had reached the ultimate glory of vision, in an audience where God had communicated with him in utmost intimacy, sitting on a throne of light. The Prophet had allegedly used the moment for a sacred bargain, by beating down the number of daily prayers from fifty to five in order to make life easier for his community. Bowing down in prayer, the parallel to the *proskynesis* in the presence of the Byzantine emperor, was in Islam the symbol of submission to God. The early Muslims called themselves *ahl al-ṣalāt* 'people of the prayer'†; they understood their way of praying as an expression of their identity as people of God, and of God alone. However, God's last prophet, by seeing Him from face to face, had brought down God's overwhelming power to human dimensions. This was more than had been achieved by Moses.

God being nothing else but Light was an idea mainly cherished by the early Shiites. God sheds His light upon his first creatures, that is, the Prophet Muḥammad and his descendants, when they are still in their pre-existence; afterwards, when the world has been created, this same light is activated in the form of a sparkle which was inserted into Adam when God shaped him in His own image. This is why Satan was wrong when he refused to prostrate himself before Adam saying that he himself consists of fire (sura 2:33, etc.). Light tops fire. God also registers everything which happens on earth by touching it with the rays He emits – a distant echo of the emission theory developed in Greek antiquity. This also implies, however, that God only becomes aware of things which exist and which can be touched; He does therefore not have any foreknowledge of the future since the future does not yet exist. The concept of predestination could therefore not come up; God does not know the things which are not yet there and which otherwise He might have predestined. A Shii theologian of the second half of the second century H. then conceded that God's rays even penetrate the ground; they are so strong that He is able to see what is under the earth. However, this same theologian found himself confronted with the objection

---

* Not in the Qur'ān but in a *hadīth* which was inspired by this passage; cf. my article *Vision and ascension. Sūrat al-Najm and its relationship with Muḥammad's mi'rāj*, in: Journal of Qur'anic Studies 1/1999/47–62.

† Or *ahl al-qibla*; cf. my *Theologie und Gesellschaft*, vol. IV, p. 680 and *Der Eine und das Andere* (1–2, Berlin 2011), vol. II, p. 1270.

that if God himself can be seen since He is Light, He must have a body and a form, for Light is not immaterial and vision deals only with objects which are shaped. The theologian agreed: God has a body, but this body is impenetrable just because it consists of light. Its form, however, does not resemble the form of a human being; God is rather shaped in regular geometrical dimensions, like a Platonic solid, that is, a globe or a cube. He is massive and therefore does not have the normal bodily functions; He rather resembles an ingot which glitters like silver. When He acts, He does so only by His rays. His light is not primarily an expression of power; it rather means brightness or lustre.*

Anthropomorphism had become a big issue. The rationalists insisted that the Prophet, although he had seen God during his nightly journey, could not have seen Him in the form of a human being; only idols are shaped like that. On the other hand, some people who knew the Book of Genesis pointed out that God had created Adam 'in His image', that is, according to the form He, God, had Himself. They liked to add that God is beautiful and that He shows His beauty when He allows a person to see Him.† However, the rationalists objected, this is only idle talk and misleading imagination.‡ There is no form (μορφή) whatsoever, and the best solution is to reject such statements altogether or, if this is not possible, to reinterpret them. When the Qur'an says that God is the 'Light of Heaven and earth', this only means that He is the lord of the universe; we are dealing with a metaphor. To be on the safe side, we better admit that God cannot be seen at all, not even in the form of light.

So sober a position was, however, again not to everybody's taste. Many people expected at least a real vision of God in Paradise; this hope is sometimes explicitly formulated on early tombstones.§ Of course, there were also other things to be hoped for in the Islamic paradise, but seeing God, that is, seeing His beauty, was considered to be the top event. People imagined it as a kind of theatre performance; God was thought to be sitting on a stage behind a curtain, and this curtain would be raised from time to time. Vision is therefore intermittent, and God remains completely inactive; the Blessed enjoy certain moments of mere contemplation where God does not emit rays at all. The visionary process thus starts from the eye of the beholder. One of the rationalists, though insisting on the impossibility of any direct vision, still conceded that the Blessed might be equipped with a special way of perception, a kind of sixth sense which will achieve what the human eye is not able to do.¶ However, one generation later, even this loophole was closed. Rationalist theology, if it allowed beatific vision at all, only permitted that man sees God 'in his heart'. Seeing Him does then not mean more than imagining Him, albeit in all His power

---

* The theologian I have been talking about is Hishām b. al-Ḥakam who died in 197/812; cf. my *Theologie und Gesellschaft*, vol. I, p. 349 ff. (especially p. 361 f.). He is said to have described God as a 'compact body of light' (*jism ṣamadī nūrī*).

† Cf. my article *Schönheit und Macht. Verborgene Ansichten des islamischen Gottesbildes*, in: E Hornung (ed.), *Schönheit und Maß* (Basel 2007), pp. 15–42.

‡ Cf. now Ch Melchert, *The Early Controversy over Whether the Prophet Saw God*; in: Arabica 62/2015/459–476.

§ Cf. W Diem and M Schöller, *The Living and the Dead in Islam* (1–3, Wiesbaden 2004), vol. I, p. 173.

¶ Cf. my *Theologie und Gesellschaft*, vol. III, p. 49 f., for the proto-Muʿtazilite Ḍirār b. ʿAmr who was a contemporary of Hishām b. al-Ḥakam.

and beauty.* In the language of the rationalists, looking at something (*naẓar*) had come to mean speculating about it.

There was no doubt, however, that as long as we live, we won't have any experience of the Hereafter anyway; we can only believe in it. Speculation may be good, but we better restrict it to what can be seen on earth. What is it then that we see when we open our eyes in our normal environment? At this point, the man who had come up with the sixth sense – he lived in the time of Hārūn al-Rashīd (or, if you want, Charlemagne) – developed an interesting theory which brings us back to Late Antiquity.† What we see is not light, he said; it is colour. When light meets or touches the surface of an object, our eye perceives the colour of this object. We do not perceive the object itself; we perceive only 'accidents' or 'parts' of it. However, our perception is not restricted to the eyes; we also perceive the noise made by an object, its smell, its taste and, besides its colour, also its dimensions, its temperature, etc. The object as such is only a construct resulting from the φαινόμενα we perceive; this is how and why we know about its existence. In the Hereafter, however, in another world, we shall come to know what the object really is, that is, its essence (*māhiyya*). Our experts of intellectual history call this a 'bundle theory', and they tell us that there was more of this in Late Antiquity.‡ Such a theory could, of course, not be applied to God; it had been developed in a world which had not yet cared about theology. However, for our Muslim author, the cognitive process now consisted of two phases: On earth, we perceive reality only with our senses, and we know about God mainly by revelation; in Paradise, however, our knowledge of God will be raised to another level by a sixth sense. We feel tempted to compare this with the famous dictum of St. Paul that on earth we see everything only 'as in a mirror', ἐν αἰνίγματι, or 'piece by piece' (ἐκ μέρους), whereas in Paradise we shall see 'face to face'(1. Cor 13.12).§ However, when our Arab author restricts vision to colours, he rather seems to be aware of Aristotle¶; as a matter of fact, he wrote something against Aristotle's doctrine of 'the substances and the accidents'.** Unfortunately, we don't have this refutation any more, and even his bundle theory is known to us merely by way of a few doxographical remarks. What we can say is only that the rationalist theology we have been talking about was decided to grapple with the affairs of the sublunar world as well as with those of Heaven, though in a

---

* For a summary of this development cf. *Theologie und Gesellschaft*, vol. IV, pp. 411–415; also Cl Gilliot, *La vision de Dieu dans l'au-delà*, in: M A Amir-Moezzi e. a. (edd.), *Pensée grecque et sagesse d'Orient. Hommage à M Tardieu* (Turnhout 2009), pp. 239–269.

† Ḍirār b. 'Amr again; cf. *Theologie und Gesellschaft*, vol. III, pp. 38–44 (where, however, the focus is not on the concept of light) and, more precisely, my article *Schauen und Sehen als ontologisches Problem in der frühen islamischen Theologie*, in: E Chaumont (ed.), *Autour du regard. Mélanges Gimaret* (Louvain 2003), pp. 1–13, especially p. 4 f.

‡ Cf. R Sorabji, *Matter, Space and Motion* (London 1988), p. 44 ff.

§ 'Piece' comes close to Ḍirār; he spoke of 'parts' (*abʿāḍ*). The 'mirror' is only found in the English translation; ἐν αἰνίγματι means simply 'in an enigmatical way'. Cf. the title of Barbara Tuchman's once famous book on the One Hundred Years' War in medieval France: *As in a Distant Mirror*.

¶ *De anima* II 7. 418a 29: τὸ γὰρ ὁρατόν ἐστι χρῶμα; cf. D C Lindberg, *Theories of Vision from al-Kindī to Kepler* (Chicago 1976), p. 7 f.

** *Radd ʿalā Arisṭālīs* (sic) *fī l-jawāhir wal-aʿrāḍ*; cf. *Theologie und Gesellschaft*, vol. V, p. 229 nr. 8.

general philosophical way and not yet with the pragmatic approach typical for Ibn al-Haytham.

The two centuries which still separate us from our hero have to be passed by quickly. In the third century H. new actors entered the scene, the first representatives of Islamic mysticism. One of them, an Iraqi again whose name was Nūrī (which is derived from the Arabic word for 'light'), imagined his mystical union with God as an experience of being consumed in light. 'One day I looked into the light', he is supposed to have said, 'and I continued doing so until I became that light myself'.* Contemplation had transformed him; in exchange for having been annihilated as a person, he participated in God's luminary essence. Light, by showing its power, had destroyed his individuality. However, he does not complain and nor does he pretend to have been victimised. God cannot be cruel, and death is a moment of exaltation and fulfilment. We are told that when Nūrī died, he heaved a sigh like in ecstasy; he had reached the moment when the deadly power of light was to be replaced by real vision.†

Mysticism was not yet a mass movement; one century before Ibn al-Haytham came to Cairo, we witness in Baghdad the public trial against Ḥallāj, a mystic who was hanged on the charge of heresy. Ibn al-Haytham may not have heard about this; the event became spectacular only in the perspective of later generations. However, we know from Ibn al-Haytham's bibliography that he was aware of what was going on in rationalist theology. He criticised its representatives (the so-called *mutakallimūn*) for the way they argued against the philosophical doctrine of the eternity of the world,‡ but also blamed the sceptics who said that all arguments are alike and that we will never reach the absolute truth (the Greek doctrine of ἰσοσθένεια τῶν λόγων).§ He had his own opinion about the way in which Rāzī, a Baghdadi physician whom we know as Rhazes in the Latin tradition,¶ attacked the Islamic concept of prophecy,** and he did not agree with the excommunication of an Iranian theologian, a man by

---

* Cf. R Gramlich, *Der eine Gott. Grundzüge der Mystik des islamischen Monotheismus* (Wiesbaden 1995), p. 339 f.

† Abū l-Ḥusayn al-Nūrī died in 295/907–908; cf. the chapter in: R Gramlich, *Alte Vorbilder des Sufitums*, vol. I (Wiesbaden 1995), pp. 381–446, especially 388 f. For the entire development cf. my article *Schauen und Sehen* (see n. †, p. 18), pp. 10–12.

‡ *Maqāla fī anna l-dalīl alladhī yastadillu bihī al-mutakallimūn ʿalā ḥudūth al-ʿālam dalīl fāsid, wal-istidlāl ʿalā ḥudūth al-ʿālam bil-burhān al-iḍṭirārī wal-qiyās al-ḥaqīqī*; cf. Heinen (see n. * on p. 14), p. 271, 4 f. > Ibn Abī Uṣaybiʿa II 97, 12 ff.). Cf. also his *Risāla fī buṭlān mā yarāhu l-mutakallimūn min anna llāh lam yazal ghayr fāʿil thumma faʿala* (265, 15 f. > IAU II 95, 6 f.), on the inediquacy of their reasoning that God was inactive during an eternity a parte ante (*azal*) and then (at a certain moment) started acting (by creating the world). Later on, between 417/1027 and 419/1029, he treated the topic again (271, 10 f. > IAU II 97, 16 f.), probably during a sojourn in Iraq.

§ *Kitāb fī n-naqd ʿalā man raʾā anna l-adilla mutakāfiʾa* (266, 17 Heinen > IAU II 95, 20 f.) For ἰσοσθένεια τῶν λόγων cf. now M Carlos Lévy, *Scepticisme et rhétorique, de Pyrrhon à Sextus Empiricus*; in: Comptes Rendus de l'Académie des Inscriptions et Belles Lettres 2012, pp. 1285–1301.

¶ Cf. EI² VIII 474–477 (L E Goodman) and DSB XI 323–326 (Sh. Pines).

** *Naqḍ ... Ibn al-Haytham ʿalā Abī Bakr al-Rāzī al-mutaṭabbib, raʾyahū fī l-ilāhīyāt wal-nubūwāt* (270, 9 f. Heinen > IAU II 97, 6); cf. also his *Kitāb fī ithbāt al-nubūwāt wa-īḍāḥ fasād raʾy alladhīna yaʿtaqidūna buṭlānahā wa-dhikr al-farq bayna l-nabī wal-mutanabbī* (270, 14 f. Heinen > IAU II 97, 8 f.).

the name of Ibn al-Rāwandī who in later centuries was considered to be an arch heretic.* He was an independent mind.†

It seems that we are able to date three of these treatises (although they are all lost); they were composed between 1027 and 1029.‡ He may have been in Baghdad then, at least for a short interval§; in Iraq, these questions were still a matter of debate.¶ However, he had certainly not left Egypt because Baghdad was more 'orthodox'. 'Orthodox' is again one of those terms which are not of much help in that period. For not only was Islam a secular civilisation, as I said before, it also was pluralist to an extent unheard of in our days. The fundamentals of Islam were only vaguely defined. Certainly, everybody believed that there was only one God and that Muḥammad had founded Islam. However, as for the rest, there was not much of a consensus. The Abbasid caliphate in Baghdad had lost its vigour; it still possessed some authority but retained no power because it could not afford an army any more. In Egypt, the Fatimids had established their countercaliphate. They were Shiis, but the majority of their subjects who lived in the old parts of Cairo, not to mention the peasants out in the country, were Sunnis (if we are allowed to use this term which was not yet used in those days). In Iraq, a princely family from Persia, the Būyids, had seized power and acted as majordomo to the caliph. They, too, had Shii leanings; therefore, the period around the year 1000 can also be regarded as a Shii interlude, both in Egypt as well as in Iraq. The Baṣra area where Ibn al-Haytham's parents had lived became the playground of an extremist Shii sect which was said to accord divine honours to its leaders, the Shābāshiyya or Shabbāsiyya. We know next to nothing about them, but their protagonists seem to have worked in the local administration.** When in Cairo the caliph al-Ḥākim was assassinated, his most extremist adherents who believed in his divinity – we know them in our days as Druzes, now in Lebanon, Syria and Israel, not in Egypt – sought contact with those people in Baṣra in order to find some place of refuge.†† Other

---

* *Maqāla ... fī īḍāḥ taqṣīr Abī 'Alī al-Jubbā'ī fī naqḍihī kutub Ibn al-Rāwandī wa-luzūmihī mā alzamahū īyāhu Ibn al-Rāwandī bi-ḥasabi uṣūlihī, wa-īḍāḥ al-ra'y alladhī lā yalzamu ma'ahū i'tirāḍāt Ibn al-Rāwandī* (270, 16 ff. Heinen > IAU II 97, 9 ff.). For the historical context cf. my *Theologie und Gesellschaft*, vol. IV, pp. 316–349.

† He was obviously also very ambitious. This is at least the impression one gets from his bibliography which he compiled himself and which reaches dimensions comparable only to what we know from the modern internet. In one of his treatises he even dealt with the difference between Greek and Arabic poetry (p. 264, 12 Heinen > IAU II 94, -7), a topic rarely touched by Muslims of this period. Can we explain his prolific output by the fact that he was short of size? Bayhaqī tells us that Ibn al-Haytham had to climb on a bench when he presented his project to the caliph al-Ḥākim who was sitting on a donkey (*Tatimma* 86, 1).

‡ That is, those which are mentioned in n. ‡, **, p. 19 and n. *, above; they are found in the second list of his books which was compiled between 417 and 419 H.

§ A stay in Baghdād for a few months is attested for the year 418 in his bibliography (271, 8 f. Heinen > IAU II 97, 15 f.).

¶ The entire second list of books, that is, his publications between 417 and 419, may reflect a move from Egypt to Iraq. The list starts with his comment about al-Rāzī which seems to have been included in a *Risāla* directed to the Christian physician Ibn al-Ṭayyib al-Baghdādī who died in 435/1043 (Heinen 270, 8 ff. > IAU II 97, 4 ff.; cf. EI² III 955). Moreover, Ibn al-Haytham seems to have communicated all the time with people he knew in Baṣra and Ahwāz (268, 5 f. Heinen > IAU II 96, 10 f.).

** Cf. EI¹ Suppl. 219 > EI² IX 159 s. v. *Shābāshiyya* (L Massignon).

†† Cf. my *Der Eine und das Andere* II 1357, with some further detail.

Iraqi Shiis inflated the theology of light by focusing on shadows (*azilla*) which were supposed to have made all sorts of mischief,* and in Egypt an early Ismaili author had come to believe that when God had created Light as His first servant in order to be relieved Himself of any further work, Light miscalculated the situation and fancied to be the Creator itself – a kind of palace revolution in Heaven.† These are gnostic speculations which are again rooted in Late Antiquity; they represent the irrational side of Greek heritage. They are not completely extinct in our days; some of them live on in the spiritual heritage of modern Syria, among the ʿAlawīs (or Nuṣayrīs). There is no doubt, however, that Ibn al-Haytham's mind was tuned to a different music.‡ He seems to have remained the sober intellectual whose concern was demythologisation, 'Entmythologisierung' as was the term in German protestantism when I was young.

Theology, wherever we find it, is a world dominated by constructs. In some of the constructs mentioned in our survey, Iraqi theologians conceived light as a kind of 'Urphänomen': Although it is true that God created the luminaries of our terrestrian environment, that is, the sun and the moon, only He himself was and is Light itself, in all eternity and already before creation. In a way, this approach amplified what we know from the Book of Genesis, namely that Light was created two 'days' before the sun and the moon. Ibn al-Haytham, however, only dealt with the light we experience every day; therefore, he was free to assume that the moon has no light of its own but

---

* I am alluding to the *K al-Haft wal-azilla*; cf. H Halm, *Die islamische Gnosis* (Zürich 1982), p. 240 ff. where larger portions of it are translated.

† Cf. H Halm, *Kosmologie und Heilslehre der frühen Ismāʿīlīya* (Wiesbaden 1978), p. 75 ff.: 'Die Verblendung des Demiurgen', referring to a *Risāla* written by Abū ʿĪsā al-Murshid, an Ismāʿīlī dāʿī who lived and acted under al-Muʿizz, that is, at the time when Cairo was founded. Similar ideas circulated among the early Druzes (Halm, ib. 80 ff.) and thus in Ibn al-Haytham's immediate environment.

‡ He may have stood in a Shii tradition though. In the preface of his autobiographical *Risāla* he praises the Prophet 'and his family (ʿitra)' (p. 258, 6 f. Heinen), which is rather a Shii way of starting a literary discourse, and his name is normally given as Abū ʿAlī al-Ḥasan b. al-Ḥasan (sometimes also 'al-Ḥasan b. al-Ḥusayn'). This is not much by way of evidence, but we have nothing of this sort with respect to a Sunni background. When Bayhaqī says that Ibn al-Haytham observed the Law and was very pious (*Tatimma* 87, 6) we have to take into account that this author had good relations with the Shiis; for he produced a work on the genealogy of the ʿAlids, the *Lubāb al-ansāb* (ed. al-Sayyid Mahdī al-Rajāʾī, 1–2, Qum 1410/1990). We might have to approach the problem from a different angle anyway. In the *Risāla* which contains Ibn al-Haytham's bibliography he is called (or calls himself) Muḥammad b. al-Ḥasan; this is therefore also the heading under which he appears in Ibn Abī Uṣaybiʿa's work. Ibn al-Qifṭī, on the contrary, knows him as al-Ḥasan b. al-Ḥasan, and 'al-Ḥasan' is, according to the common opinion, also the *Vorlage* for Latin 'Alhacen' (or Alhazen). Rushdi Rashed has tried to solve the conundrum by assuming that there were two Ibn al-Haythams, one of them living in Egypt under the name of al-Ḥasan and the second one living in Baghdad one century later under the name of Muḥammad, that is, the man who combined the two lists of Ibn al-Haytham's books in 556/1161; cf. *Encyclopedia of the History of Science* (see n. †, p. 14), p. 1090a. This may be sustained by the fact that there is, apart from the two lists, a third one which was made for Ibn al-Haytham's son in law (*khatan*, if the reading is correct) and where he is again called 'al-Ḥasan' (Heinen 273 ff. > IAU II 97, -10 ff.); it covers the years until 429 H., that is, until shortly before Ibn al-Haytham's death. On the other hand the Latin form *Alhacen* could also be derived from 'al-Haytham' instead of 'al-Ḥasan'; in comparison to the *shuhra* the *ism* was, after all, not so commonly known. Unfortunately al-Baihaqī has no *ism* at all; he simply says 'Abū ʿAlī Ibn al-Haytham'. It might be helpful to have recourse to the manuscripts of Ibn al-Haytham's works in order to solve the problem.

owes it entirely to the sun.* Beyond that, he shunned any mythmaking and preferred the *clarté* of Greek antiquity: Aristotle, Galen, Apollonios of Perga, Archimedes and their like. What remains to be discussed is whether myths ever die; they rather seem only to take on a new apparel. When we, in our generation, managed to create a light which was 'brighter than a thousand suns', we evoked new illusions and brought to life new heroes. Instead of God, we got acquainted with a 'terminator' who entered the mythology of our films and became so popular that his admirers elected him as their governor, and we ultimately found ourselves surrounded by hordes of 'shadows', secret agents behind the scene who were ready to kill without any warning. Light became a weapon, and we thought that it was our human right to keep it for ourselves. Some further 'Entmythologisierung' might be in order. The year of light is over, and enlightenment still far away. We have been looking back to Ibn al-Haytham across one millennium, and we think that we have made some progress. However, will we have another millennium to survive?

---

* He did so in the treatise mentioned above (see n. ‡‡, p. 12) as being copied by Ibn Riḍwān during Ibn al-Haytham's lifetime. The text is preserved and was translated into German by Karl Kohl in: Sitzungsberichte der Physikalisch-medizinischen Sozietät in Erlangen 56–57/1924–1925/305–398. Cf. Schramm, *Ibn al-Haiṯams Weg zur Physik* 146 ff. and Sabra in DSB VI 195 and 205 nr. III 6.

3 From the Reasons of Light to the Lights of Reason

Remarks on the Nine Centuries Distant Ibn al-Haytham's and Albert Einstein's Respective Approaches of Light as Conceived Physically*

Michel Paty

CONTENTS

Introduction – Prologue: Light As 'Physical' and Rationality 23
Ibn al-Haytham's Physical Optics As Fundamental Basis for Geometrical Optics 25
Einstein's Two Fundamental Contributions in Taking Full Account of
    the Physical Nature of Light, the Special Relativity Theory and the
    Quantum Approach...30
Epilogue: Physics and Rationality – The Requirement of Objectivity and the
    Conditions for Mathematisation 36
References...38

INTRODUCTION – PROLOGUE: LIGHT AS 'PHYSICAL' AND RATIONALITY

The metaphor of light as illuminating to symbolise the understanding in knowledge according to reason is so commonly used that it permeates the language, probably in all cultures, for it refers to an immemorial experience in daily life. Think, for

---

* Contribution to the Unesco Colloquium 'The Islamic Golden Age of Science for today's knowledge-based society: The Ibn Al-Haytham example', 14–15 September 2015, Unesco, Paris, France.

example, of Plato's allegory of the Caverna proposed in his *Republic*,* where one sees only shadows of forms that contrast with the full vision of bodies when one has got outside, in the daylight.

Beyond the image, still actual, the history of scientific and philosophical thought confirms the close relationship between being illuminated by *light* considered as *'physical'*, with the general meaning of pertaining to the natural world, and getting the understanding from the use of *reason* or *rationality*, particularly in the development of science (including the science of light that Optics is); and also, more generally, in the development of ideas, particularly philosophical ideas, that go along and interact with the first one, under the banner of the 'lights of reason', as it was currently expressed after Descartes and especially in the eighteenth century, which was by excellence the 'age of the Enlightenment'.

We shall evoke, in this perspective, two significant moments in the elaboration of the science of light, situated in time nine centuries one from the other (at both ends, so to speak, of the history of the knowledge of the *physical constitution* of light), corresponding to the innovative works, respectively, of Ibn al-Haytham and of Albert Einstein.

Between the two stands a series in correlation of other fundamental works that we shall only be able here to mention without any details (from Kepler to Descartes, Fermat, Grimaldi, to Newton, Huygens, to Young, Fresnel, Maxwell, Lorentz, among a number of important authors). We would indeed have, with these, other significant examples of researches quite illustrative, each one in its own time, of the relationship between, on the one side, fundamental, structural discoveries on light considered as physical and, on the other side, the movement of the rational thought that endeavours to describe and to understand, and thereby operates its own overcoming, enlarging the forms of rationality, physical and mathematical rationality, in the case of light that we consider here.

The same applies to forms of rationality at work in other sciences or in other areas of the human experience, although with the peculiarity, in the case of light as well as of other objects of the physical science (this time taken in the more restrictive meaning that Physics has nowadays as one specific among the other natural sciences), that the proper rationalisation of it corresponds to the setting up of the conditions for a mathematised physical theory.

Such will be, as an effect, the philosophical lesson to come out from the two cases of pioneer elucidation that we have chosen to evoke, taken in the history of science, with no direct connection between them (if not that one of a very distant filiation), and in extremely different intellectual and social contexts. This lesson bears, for each case, on the nature and operating modes of the *human thought* in its quest for scientific knowledge, dealing and struggling with the *phenomena* of the material world, such as these are given in the experience of the actual world through the senses and the mind, and trying to conceive them as if 'from outside', that is, according to *objectivity*, and to get the *understanding* of them.

It is indeed such a project, defined in these relatively general terms, that can be found under well different characteristics and knowledge contents, in the works of these two scholars that nine centuries separate. We shall be careful, of course, to guard against any anachronism or inadequate comparison and approximation.

---

* Plato's *Republic*, Book seven.

However, this worry does not forbid one from seeking meaningful connections, and even traces of invariants from one to the other in the formulations and achievements of their respective endeavours.* This is obviously important if one wants to keep the idea of *unity of the human reason*, under various forms, temporal as well as regional, of rationality.† We will see in them, as for us, two illustrations, expressed in different words and claims, but of neighbouring scopes, of the path of reason according to the specificity of the object of its attention. From there springs out the proper meaning, in each case, of the 'light of reason', as well as the general scope that can be attached to this expression.

Owing to the volume limitation of the present contribution, we shall evoke first Ibn al-Haytham's works in Optics and then jump over nine centuries (so rich as they were!) directly to Einstein's ones, assuming, however, that the reader keeps in mind at least some of the most original acquisitions in the intermediate period that have given significantly more precision to the *physical nature* and *properties* of light, while elaborating correlatively, through various phases, a *physical* (with variable meanings) *theory* of it that was *mathematically expressed*.

Strictly speaking, and with the present meaning of the qualification of '*physical*', it is only with the Electromagnetic Theory of Maxwell, and then with the Quantum Theory (more specifically, with Quantum Electrodynamics), that the theory of light is considered as fully physical, in that it describes with precision *what physically light itself is*, this description determining as a consequence its properties (light is an *electromagnetic wave* with the first, a *state of the quantum electromagnetic field* with the second). In the nineteenth century, before Maxwell's theory, one preferred to speak of the Wave Theory of Fresnel as the 'Mathematical Theory of Light',‡ in the absence of a definition for light more precise than the propagation of a 'disturbance of the ether', this latter (the optical ether) being thought according to Mechanics as a kind of elastic solid medium filling space: this idea, however, was very useful as it allowed one to describe its properties by making use of the Mathematical Theory of Partial Differential Equations. It is with a somewhat similar perspective (modulated by the cultural shift of the historical time) that Ibn al-Haytham significantly conceived and developed in his time the geometrisation of an Optics that he was seeing as 'physical', in an even less precise meaning, for sure, but nevertheless justified, considering the intention and the deep significance, as we shall see now.

IBN AL-HAYTHAM'S PHYSICAL OPTICS AS FUNDAMENTAL BASIS FOR GEOMETRICAL OPTICS

Ibn al-Haytham (end of the tenth and first half of the eleventh century, born in present-day Iraq), later known in the West as Alhazen, is the author (among other

---

* Paty (1999a, b, 2007).
† We take the expression 'regional rationality' with the meaning Bachelard gave to it: Bachelard (1949).
‡ As, for instance, Poincaré entitled his first books of lessons on light, *Mathematical Theories of Light* (*Théories mathématiques de la lumière*, Poincaré (1889–1891)), whose scope was voluntarily limited to the (mechanical) Ether Theories of Light, while he published soon after the exposition of the Electromagnetic Theories of light in a second series of two volumes under the title *Electricity and Optics* (*Electricité et Optique*, Poincaré (1890–1891) 1901).

fundamental works in Physics, Geometry and Astronomy) of a treatise, *The Book of Optics*, that would come later (six centuries later) in a Latin translation to Johannes Kepler, inspiring him in his researches in Optics, linked to his works in Astronomy. It was long believed that the work of this Arab scholar was isolated in the scientific tradition dating back to Claudius Ptolemy (second century) and leading to Kepler (end of the sixteenth – beginning of the seventeenth century*). However, this simplistic view has been radically questioned thanks to the recent fast development of research on the history of Arabic (or Arab-Islamic) science, through which important texts of scholars and thinkers of this cultural area, hitherto unknown even from the specialists, have been rediscovered and identified, revealing a rich tradition of research throughout centuries (from the ninth to the fifteenth centuries) in various fields of scientific knowledge, and that made the link with the Greco-Latin culture[†] (and also with other Oriental cultures, such as that of India). Part of this legacy had reached, through latin translations, the West in the Christian Middle Age, thus contributing to inspire thinkers, adding to that one received directly, essentially from Byzance. However, many original contributions had been lost and forgotten meanwhile. The mentioned rediscovered ones allow now to shed light even on those contributions that were thought to be well known, either by adding more texts from a given author that complete (be it partially) the previous ones, either by the possibility to better situate them in their (intellectual) contexts.[‡]

Such is the case in particular with the works on Optics of Ibn al-Haytham, which were widely but only partially known, and that have been translated, scientifically edited and analysed, in particular by Roshdi Rashed, who has also been able to re-situate them in a series of works done by predecessors and successors, which he has revealed, allowing on the overall a more complete account of what exactly was known as for the science of Optics at that time, and how these various authors did think it.[§]

Ibn al-Haytham's overall work is impressive by its volume, by the extension of its fields of interest, by its scope that includes many subjects and that joins theory to experiment, and also by its richness as to the new results obtained, manifesting a high degree of inventivity, always acompanied by methodological and philosophical reflection. As to his Optics, to which we shall limit ourselves here, even if the importance of his contribution was recognized very early, we dispose now of a better informed and more comprehensive knowledge of his researches, put in relation with those of

---

* Before Kepler, Ibn al-Haytham's optical works were known, also in translations by Roger Bacon and Vitello (thirteenth century). See, in part, Lindberg (1997).

† The translation in Arabic language of the Greek and Latin texts was essentially achieved in the first half of the ninth century.

‡ I owe most of my (limited) knowledge of the Middle Age sciences (Mathematics, Optics and Astronomy) in Arabic language from reading the work of R Rashed and his disciples and collaborators. I am grateful to them, and I advocate their indulgence towards the neophyte. My interest in this recent chapter (despite the seniority of its object) of the history of science sometimes led me to epistemological reflections: cf., for example, Paty (1985, 1987, 2007), and also on more general aspects: Paty (1999a, b).

§ Let us mention, among others, writings by Ibn Luqa, al-Kindi, Ibn Sahl (tenth century), and several ones by Ibn al Haytham (eleventh century) that were lost and have been recently discovered. See Rashed (1997).

other scholars of that time, which allows a more precise evaluation of the nature of the reform he has proposed, and a full appreciation of its meaning and scope.

We cannot omit first to evoke, even very briefly, among his direct predecessors, the researches of Ibn Sahl (second half of the tenth century),* author of a considerable work, notably in Geometry, and who proposed original results in Optics, that he treated according to Geometry, for example by using the properties of the conics curves to describe the path of the light reflected by mirrors or refracted by lenses. Ibn Sahl's interest for these questions arose from his project to understand burning mirrors (known since Archimedes, and whose study was part of Catoptrics). He was the first to conceive the extension of such phenomena to lenses of various types, produced by refraction of parallel light rays, that could, depending on the lens form, be focused to a point of burning. His studies on the lenses performed in this perspective opened the development of the new science of Dioptrics.

Ibn Sahl, who was much more a mathematician and a geometer than a physicist (despite his strong interest in the phenomena of light-induced burning), established many important results (including a form of the law of refraction later called 'of Snell-Descartes'), but without regard to the physical nature of light, and with no interest whatsoever in the doctrine generally adopted since the Antiquity, of the 'visual ray'. However, despite his lack of commitment to Physics itself, he formulated new notions that have to do with considerations about Nature: such as the notion of 'medium' (of the propagation of light), of larger or smaller 'transparency' and 'opacity' of it, and of a constant ratio for each medium in the deviation by refraction. These concepts would be taken in his turn by Ibn al-Haytham with a more definite physical concern, and given more precision (they would be, later on, at the origin of the concept of refractive index).

Ibn al-Haytham was aware of the work of Ibn Sahl when he undertook his own researches, and was certainly highly receptive to his predecessor's geometrical results.[†] Being himself a mathematician, as Ibn Sahl, but as well also, differently from him, a physicist, he had, as for him, a strong concern for the 'physical' (in the sense of 'given in nature') character of light, and was eager to know its properties, namely those of propagation in different media and conditions, by *'composing mathematics and physics'* with respect to it, as it was done for bodies submitted to gravity.

However, the traditional received conception, since Euclid and Ptolemy, was that of the *'visual ray'*, which puts in the foreground *vision*, conceived as the emission by the eye of *visual rays* (the nature of which was let undefined) that hit the material objects (the 'visibles') whose images are sent back to the eye in the visual cone, either as a whole (conception known as 'intromission of forms') or by emission of light by its different parts ('emission' conception, professed by Euclid).

---

* See especially Ibn Sahl's *Treatise of burning instruments* and *Evidence that the celestial sphere is extremely transparent*, see the texts (Arabic original and French translation) in Rashed (1993), pp. 1–82 (double pages); and the studies on these texts given in the same volume, Chapters 1 and 4.

† On the work of Ibn al Haytham in Optics, see especially the translations made by A I Sabra and R Rashed, and among the available studies: Rashed (1978, 1992, 1993, 1997) pp. 309–318. As R Rashed shows in his comparative approach of the two scholars, Ibn al Haytham did not adopt in his own work on refraction the law of Ibn-Sahl(-Snell), for reasons that have to do with his own project.

According to this conception, Optics was a kind of a 'science of vision' or a 'geometry of perception', in the words of Gérard Simon,* that prevented a full account of light in the physical sense and opposed any significant development of Optics. It is only through a *physical* conception of light, that is to say, admitting the *materiality* of light rays (even taken only as a matter of principle, without knowing more details about its nature), that the rectilinear propagation and the geometrical behaviour of light rays could be really justified. The doctrine of the 'visual ray' was in fact an 'epistemological obstacle'† to the development of a true *science of light*, even if it had accommodated the rectilinear propagation and geometrical properties studied by Catoptrics.

It is precisely this conception that Ibn al-Haytham rejected, and this questioning constituted the turning point in his reform of Optics. His reform consisted primarily in rejecting the traditional doctrine, accepted since Antiquity, of the 'visual ray'. He considered light no more as an emanation from the eye, but as a *natural entity* (in his aristotelian vocabulary, a 'substantial, or accidental quality'), that bodies emit (in the case of the luminous bodies) or return (in the case of illuminated bodies), and that is propagated from them in space, in all directions, eventually towards the eye if this one happens to be properly located. By posing the *physical character of light*, he separated the *theory of light* and *that of vision*, the first being the condition of the second (which, moreover, he also studied‡) and not the reverse, overturning the view previously admitted.

The conception of light as existing independently of and prior to vision and to the subject (of knowledge) opened the possibility of a *Physical* (meaning *Natural*) *Optics* which would be understood henceforth as the *science of light and of its properties*, from a perspective of objectivity. This was a refoundation of Optics as a science, for Ibn al-Haytham studied Optics and not only (as Ibn Sahl did) mirrors and lenses, and it was possible for him to define in all generality and precisely the *object* of that science. Such indeed was what changed from one to the other of these authors, and as well in comparison with the traditional view: the very *object* of the science addressed at. And, differently from his predecessor, who admitted *a priori*, as by definition, the geometrical properties of light rays, he founded them, as for him, on physical reasons.

In his *Book of Optics*, Ibn al-Haytham proposes to 'compose Mathematics and Physics' for the sublunar world by first highlighting the *physical conditions* required to *process geometrically* light rays. The physical nature of light (of the 'illuminant agent') as support of its propagation in space appears as the prerequisite, even without being able to say much more about it, if not in qualifying light as matter of *'fire'*, as he used to do by invoking the phenomenon or experiment of the ignition of objects on which a light beam is concentrated. He was concerned with the *physical* characterisation of the properties of light by *concepts*, based on *experiments* he himself practised: such was the concept of *light ray* that he proposed as 'the smallest light

---

* Simon (1988), p. 187 sq., quoted in Rashed (1997).
† In Bachelard's sense. See, in particular, Bachelard (1938).
‡ It is by starting from vision that the knowledge of objects (and indeed also that one of light itself) will be elaborated, and Ibn al Haytham himself devoted part of his research outlined in his *Book of Optics* to a description of vision by means of the eye, connected to the brain. See Russell (1997).

element', an obviously qualitative formulation, but that was suggested by the experiment performed with a darkroom provided with a small slit, where a thin ray entering formed a tiny light spot on the opposite wall. He also recurred to experiment with light rays outgoing from a small aperture, performing a sighting with a ruler along the light ray, from which he formulated the rectilinear propagation of light, on which Geometrical Optics relies, 'for the first time (…) in all generality'.*

Through a research work that was both experimental and theoretical, in which rational analysis plays a fundamental role, and making use of concepts such as *opaque* or *transparent bodies*, as well as other ones, and by implementing Geometry, Ibn al-Haytham was led to formulate properties that hold to the materiality of light and that constitute the characters of its propagation, of its reflection by opaque bodies and its refraction in transparent ones, considering various media. Note that, for Ibn al-Haytham, experiment is not reduced to its empirical aspects, but it is also a part of the 'proof process' and participates therefore in the work of the theoretical thought.

By taking light as the object and the physical support of the light ray, of its propagation, and of the latter's laws, Ibn al-Haytham was thinking of it, actually, as he was representing to himself bodies in motion, despite the difference evident between light and weighty bodies, that made the analogy rather distant and blurred. He interpreted anyhow in such a way the reflection and refraction of light by analogy with the changes of motions in the shocks of colliding bodies, and he might have conceived thereby the introduction and use of Mathematics in Optics considered as physical that way. This rapprochement, more intuitive than strictly justified, but fructiferous, of light and material bodies, of Optics and Mechanics, and the role of Geometry for the one as for the other, would accompany throughout the following centuries, with various peripetias and, notwithstanding its still intuitive character, getting better founded and more effective justifications (e.g., in Descartes and Newton), the history of the progressive mathematisation of Physics.†

To sum up, it is the formulation, argued rationally and on an experimental basis, of the *physical* nature of light, conceived as independent of vision, which allowed Ibn al-Haytham to propose a new and original conception of Optics, modifying the traditional status of that science, and setting rationally the means of the approach to it. According to this conception, the science of Optics is to be considered simultaneously as *physical* and as *geometrical*, while maintaining the basic distinction between the two ways of thinking (and the two corresponding scientific disciplines). Its object, light and its properties, to which luminous phenomena are referred, belongs to Nature and constitutes the material support of the science that describes it with the means of Geometry. It is, so to speak, *Optics* conceived as *Physical* that establishes the relevance and ensures the legitimacy of *Geometrical Optics*. In this perspective, the relationship between Optics and Geometry is no longer one of an ideal identification or synthesis (which would imply a basic or 'ontological' homogeneity), as it

---

* In the words of R Rashed.

† Only with Atomic and Quantum Physics in the twentieth century, in a quite different experimental and theoretical context, the close connection between light and material (quantum) bodies would be imposed by the physical facts themselves, made intelligible by Quantum Theory (see further on, in the evocation of Einstein's researches).

was conceived before this scientist, but it must be understood as 'an isomorphism of structure' between both, in the words of R Rashed.

One could hardly overestimate the importance of the innovation introduced by Ibn al-Haytham in the science of Optics. It represents a true revolution in the thought of light and of Optics, beyond the mere refoundation and the reformulation of the latter as a physical science, and the *change of perspective* it introduced could well deserve to be qualified, considering its implications, as 'Copernican', so that one could rightly speak of it as a kind of a 'Copernican revolution' (before the word), at least regarding the conception of light. Actually, we find in it the same decentration with respect to man, to the observer or the subject of knowledge, with the distanciation from the object under study taken as autonomous, and also the argumented submission to laws formulated mathematically, as well as the corresponding opening to scientific thought of a new and vast domain (although not so huge as the Universe itself as it was with Copernicus), with countless consequences thereof.... It was, however, a revolution with time-delayed effects, that would only be taken again four centuries later by al-Farisi who commented at length on the writings of Ibn al-Haytham; a Latin translation was given soon afterwards in the Christian West (with the author's name turned into Alhazen) and spread, but it would take several centuries more for it to fully bear its effects by contributing to the major changes in ideas of the Scientific Revolution of the sixteenth–seventeenth centuries.

The commentary that one can propose of these works of Ibn al-Haytham from the point of view of the question of rationality imposes itself. Beyond the elimination of an 'epistemological obstacle' (which can only be judged *a posteriori*) one observes, considering this scientist's thought in action and the ways it operated, a broadening of perspective and a deepening of intelligibility, that let emerge, formulated as such, an object of science in its materiality and its autonomy, and at the same time in its ability to be studied and described, including that to be conceived according to Geometry, as one does it for bodies by their forms and figures, while however ignoring what exactly their *physical nature* is. In other words, a new form of rationality happens to be effective, which makes possible what was not earlier, namely the thought of a *Physical Optics*, through the materiality of the beam of light rays and of the properties of the media and bodies associated with it: a *Physical Optics* concomitant with a *Geometrical Optics*, of which the first one appears as the material counterpart that guarantees the full *physical meaning* of the second.

EINSTEIN'S TWO FUNDAMENTAL CONTRIBUTIONS IN TAKING FULL ACCOUNT OF THE PHYSICAL NATURE OF LIGHT, THE SPECIAL RELATIVITY THEORY AND THE QUANTUM APPROACH

In 1905, Einstein got at the culminating stage of the formulation of the classical theory of material light that Maxwell Electromagnetic Theory was, with his Special Theory of Relativity (named so afterwards). With it, he completed the rich and significant series of works performed by predecessors in the nineteenth century (mainly from Augustin Fresnel to James Clerk Maxwell and Hendryk Antoon Lorentz) by establishing the invariance, under the (inertial) motions of bodies, of the laws of

Electromagnetism, as it was already the case for those of Mechanics. H Lorentz and Henri Poincaré reached the same conclusion at about the same time but by a different route. Most of the physicists believed at the time that the two approaches were equivalent since they led to the same result. However, the difference between both approaches is important and it would only be identified later. Let us indicate it here in its essential features.*

Einstein's theory proposed a reform that was at the same time of the physical concepts and of the theoretical structure, regarding both Mechanics and Electromagnetism, and therefore from the outset a theoretical reformulation and even a refoundation, fully aware and motivated by fundamental *physical* reasons (which does not mean empirical ones, but inseparably granted to experiments and to theoretical reasoning). It was, indeed, *to make the existing theories more physical* that he reformulated them, by redefining the magnitudes that express the concepts of space and time so that they be *physical*, and therefore homogeneous to the material bodies that are immersed in them. For they were not such up to then, in their received classical form, considered untouchable since Newton had erected these concepts as absolute ('absolute and mathematics, independent of bodies', as he wrote in his *Principia*[†]), that is to say, deliberately non-physical.

For their part, Lorentz's and Poincare's contributions[‡] were conceived by them as a mere adjustment of Maxwell's Electromagnetic Theory (already somewhat modified in 1892–1895 by Lorentz to take into account the recent knowledge on electrically charged particles, as well as the effect of motion), so as to make it compatible with a recalcitrant precise ('second order') experimental result (that of the Michelson–Morley experiment). It was therefore a rather empirical reform, yet not only so since the fitting affected the formulas of transformation of the spatial coordinates and time: the new ones, called 'Lorentz' transformations' by Poincaré, were quite different from the Galilean ones, and implied potentially, so to speak, further systematic reformulations of the space and time variables as physical concepts. However, as they were presented,[§] the modifications were still only 'touch-ups' of quantities considered as 'effective' and as limited to electromagnetic phenomena; they could, according to Lorentz and Poincaré, coexist with those, left unchanged, of Mechanics.

Einstein's theory has greater consistency from both points of views of the concern for concepts and for the theoretical structuration: it put at the centre of the considered physical theories (Mechanics and Electromagnetism) two physical *principles* (whose function is to organise the phenomena and to put constraints on the meaning of concepts and on the possible values of their magnitudes and that constitute, so to speak, the backbone of a 'principle' physical theory). These two principles were *the principle of relativity of inertial motions* (worn by all Classical Mechanics), and the

---

* For a detailed study of Einstein's proper way and a comparison with Lorentz's and Poincaré's ones, see Paty (1993), Chapters 2–4.

[†] Newton, I (1687). Galileo and Descartes who developed classical relations before Newton conceived space and time as relative, not absolute.

[‡] Proposed by texts from each of them answering to the other, and that completed themselves on the whole.

[§] On the evolution of Poincaré's thought afterwards, see Paty (1996, 2002).

constancy of the velocity of light in vacuum independently of its source's motion, stated by Einstein from Maxwell's Electromagnetic Theory. They corresponded, each one in its way, to Einstein's conception of *physical principles* (and that those of Thermodynamics provided to him the model), as general facts found in all the relevant phenomena (here, mechanical, optical and electromagnetical ones), raised in reason (and henceforth non-empirical in the narrow sense). This was not obvious at that time for the second one (most of the physicists expected a variation of the velocity of light with motion, high precision experiments being designed for this purpose), and if Einstein formulated it as he did, the reason is that he considered the value of the speed of light in vacuum, as it stood in Maxwell's Theory, as the core property of Electromagnetic Theory (that imposed itself throughout its area) which, according to him, should be maintained in any theoretical modification: he decided to give it the status of a *physical principle*.

Einstein soon realised that these two statements (these two physical principles according to him) were incompatible for classical Physics, and that therefore the two theories of Mechanics and Electromagnetism were in conflict in their present formulation; he realised that the difficulty arose from the *non-physical* character of the concepts of space and time as they were received, and in particular from the form of the Galilean law of the addition of velocities. Consequently, the two principles would not be any more in contradiction if one abandoned the classical relations and, in particular, the law of velocities; and the physical magnitudes of *space* and *time* would be afforded a proper physical meaning by being submitted to these claimed two physical principles. Einstein thus obtained, deductively, in all generality (without any additional assumption), the transformation formulas under inertial (i.e., uniform and rectilinear) motions (Lorentz's formulas of transformations), which express the invariance of the laws of physics under such motions and, in particular, a new (relativistic) formula of composition of velocities.

Space and time lost, by being physical, their mutual independence, to form a new entity, *space-time*, in which they are undissociable, where the speed of light in vacuum has (from the start) the status of an invariant and would be qualified some time later as the 'constant of structure of space-time'. These concepts or magnitudes earned in the reformulation a richer and more constraining meaning than what they had in their classical acception : the Space-Time Geometry provided a simplification and a unification in the formal description of physical phenomena and in the expression of problems, similarly to what Geometry does for space. However, as Einstein was fully conscious,* this formal idealisation cannot erase the *irreducibly physical nature* of these magnitudes, which is shown with evidence particularly in the proper meaning of time heterogeneous to space.†

However, with the Special Theory of Relativity, the physical character of space-time was still only partial; it was effective in the mutual dependence of space and

---

* Differently from his ancient professor, Alexander Minkowski, who formulated, in 1908, the four-dimension Space-Time Geometry, taking space and time as they were redefined in Einstein's theory: he used to speak of it as a 'World Geometry', but interpreting it as purely mathematical (in a Platonic sense).
† In particular, time is directional (events occur in it from the past towards the future), differently from space: see, for instance, Paty (1994, 2001c).

time but, as Einstein did put very soon after, space-time itself as a whole is independent of bodies.* It would belong to the Theory of General Relativity to overcome this 'unsuficiency'.

It has been said a few lines above that the results obtained independently by Einstein on the one side, and by Lorentz and Poincaré on the other, represented the accomplishment of Maxwell's Electromagnetic Theory (which established light as an electromagnetic wave). However, Einstein's theory (the so-called Special Relativity) was not a final outcome; rather, it was a starting point towards the *Relativistic Theory of Gravitation*, or *General Relativity Theory*.

Here again, and in an even much more decisive way, the imposition of physical principles appropriately formulated according to the considered phenomena (namely, the *principle of relativity* generalised to any kind of motion, and the *principle of equivalence of inertial mass and gravitational mass*) allowed to formulate jointly and conversely the space-time structure and the shape of the gravitational field in every point of space-time. The General Theory of Relativity was finalising the *physical character of space-time* (thought of as a continuum) which the Special Theory of Relativity had insufficiently ensured. Later, Relativistic Cosmology would reinforce this character by submitting to it the Universe conceived as a whole...

In the same period when he was elaborating these theories (in which physical concepts are expressed by continuous magnitudes or variables), Einstein found himself confronted with other problems also set by the *physical nature of light*, but bearing on another area of phenomena in which *fundamental discontinuities* appeared, at variance from the continuous processes and variables we just talked about. These were essentially the phenomena of *emission* and *absorption of light* by material bodies *at the atomic level*, which was by then a new field of investigation for Physics. (Indeed, it is at this level that light appears and disappears, is created and destroyed, through its interactions with atomic matter.)

By combining both the analyses of experiments and of the available theoretical resources, Einstein soon diagnosticated the inadequacy in this domain of the Classical Electromagnetic Theory he had contributed to complete, as we have seen. From now on, he began, and would continue unceasingly, to characterise and question critically the new, unprecedented *physical properties* of light that were being manifested and that challenged the classical theory, as well as those of the elements of matter which absorb it or produce it. He highlighted in that way the indisputable facts contrasted with the theoretical limitations, so as to specify as precisely as possible the relations of phenomena that a new theory, different from the classical ones (Electromagnetism and Mechanics) and still out of reach, should satisfy.

This exploration of a domain of material reality *beyond the direct grasp of the senses* (the domain of matter and radiation at the atomic, quantum, level) did set unprecedented challenges to the exercise of physical thought. We shall not follow

---

* Let us note, however, that the relationship of the coordinates of space and time, given in the 'quasi-Euclidean' metric of the space-time of Special Relativity ($ds^2 = dx^2 + dy^2 + dz^2 - c^2 dt^2$), is required by the physical properties of the bodies that are contained in it, and in this sense it is not totally independent of them. However, space-time and bodies are not otherwise tied; their relationship is that (external) between container and content.

here in detail Einstein's own path in these elucidations[*]; instead, we shall content ourselves with indicating summarily what is owed to him in this direction, concerning mainly light and radiation in quantum conditions, letting aside the ways of reasoning by which he established his results in this respect.

He showed first the discrete nature of the energy of radiation itself ($E = hv$, 1905[†]), and not only of the energy exchanges of radiation with atoms, which had been established by Max Planck in 1900. Correlatively, he realised shortly after (in 1906) that this discontinuity in the energy of radiation itself was incompatible with the Electromagnetic theory, and that this would be determinant for the orientation of his line of research in this field up to 1925 (we shall come back to it in what follows). Then he showed the necessity to quantise the atomic energy levels, inferring it from the 'anormal' behaviour of the specific heats of some atoms near the absolute zero of temperature as previously observed (1907; later, in 1913, Niels Bohr, on the basis of the 1911 Ernest Rutherford nuclear model of the atom duly modified into his own quantum planetary atomic model, would formulate the quantum rules of transitions between the atomic energy levels, in the direction intuited earlier by Einstein).

He was also led to propose the double (or dual) character of a wave and of a particle for light, although these characters are mutually exclusive according to the classical concepts and theories (1909–1915). Further, he elaborated a synthetic and coherent approach of a 'First Quantum Theory' (named so afterwards) by integrating together 'phenomenologically' all the quantum features of atoms and of radiation collected up to then through the works of physicists, himself as well as other ones. This synthesis (proposed at the end of 1915) was fruitful, yielding predictions for new phenomena, such as the attribution to the energy quantum of light of an impulsion ($p = h/\lambda$); and the 'stimulated emission' of radiation, ancestor of the 'laser effect'.

It can be said that this theoretical synthesis, in which he minimised the use of classical concepts relations, has been afterward the point of departure of all the further theoretical attempts that would converge to the independent formulations of Wave Mechanics and of Quantum Mechanics (in 1926–1927)[‡] (that were shown by Erwin Schrödinger to be equivalent with respect to phenomena). Last but not the least, Einstein stated the indistinguishibility of identical states of (electromagnetic) radiation as well as of atoms of neighbouring properties, later called 'bosons',[§] both exhibiting a new kind of probabilistic behaviour, known as Bose–Einstein statistics

---

* Most of the pioneer contributions of Einstein to the early Quantum Theory are included in the already published 13 volumes of his *Collected Papers* (up to 1923), see Einstein (1987–2013). A detailed study of Einstein's contributions in the quantum domain, and of his 'style' as a physicist when dealing with it, is proposed in Paty (forthcoming).

† E: energy, and v: frequency, of the light ray; h: Planck constant, introduced by him in 1900 when he stated that the energy exchanges between atom and light are integer multiples of hv ($\Delta E = n\,hv$), with n: integer number.

‡ On this, see in particular Pais (1979). On the history of these beginnings of Quantum Physics, see, for example, Jammer (1966).

§ Later on, one would add quantum elementary particles with, as such atoms and the photon of light, an integer spin or intrinsic angular momentum (more precisely, integer multiple of the quantity unit of the quantum angular momentum, $h/2\pi$).

(1924–1925).* Such were, with a few more proposed by other scientists, the fundamental properties of light and radiation and of elementary matter from which the basic Quantum Theory (under the species of Wave or Quantum Mechanics) would soon be formulated and thereafter firmly settled, throughout the years.

It is remarkable that Einstein adopted, when dealing with the quantum domain, a line of reasoning quite different from that of his works on Relativity Theories that bear on continuous fields and magnitudes. The deep reason of the difference is to be found in his conception of the strong relationship between a physical theory and the characterisation of the 'object ' it deals with. As he expressed it in one of his epistemological reflections, the continuous field theories can be formulated as 'principles theories ', for which a few fundamental physical principles are structuring and condensing all the main physical properties, as in the case of Thermodynamics and of the Special and General Theories of Relativity (and later on of his attempts at a Unified Theory of the continuous fields). Other theories are 'constructive ones', and their formulation is obtained through a more step-by-step procedure, starting from properties recorded from rationally controlled experiments.

With the quantum domain, there were no *a priori* indication from a theoretical point of view, for there was no more theory (as the classical ones failed), and even no clear specific concept available, with the exception of a few basic ones, such as energy, frequency, impulsion, position, but their physical meaning, their forms and their relationships could well be no more the classical ones. Einstein therefore decided to try characterising kinds of specific relations that can be obtained from rather simple considerations on relevant phenomena of the quantum domain. He had for this an original tool of his own, crafted from the second law of Thermodynamics (the entropy formula), to calculate fluctuations in the distributions of variables relative to the physical systems under study. Note that here also, as above, it is the *physical meaning* of such variables (and, here, of their fluctuations) that served as a guide for establishing adequate relationships. He hoped that, once some fundamental properties would be settled in that way, these would serve as an indication of phenomenal physical constraints for the future Quantum Theory (totally independent from the classical ones) that it would have to obey.

Einstein followed closely the further development of the proposed theory of Wave or Quantum Mechanics in a critical way, given the difficulties in understanding exactly what it meant physically, and the 'interpretations' that had been proposed of it, which seemed to him to deviate from the purpose he himself assigned to the theory he was looking for. In his view, it would not be a classical theory, as he wrongly has been charged, but a fully physical theory of its own, that is to say, a theory that gives account of the new properties, which departed from the usual way of thinking. He opposed, precisely, the invocation of an unescapable need for classical concepts that should be interpreted, as Niels Bohr proposed with his operationalist interpretation known as 'complementarity'.

---

* From the Indian physicist Satyendra Nath Bose, who contributed to a first stage of the elaboration of these ideas. In parallel, another kind of probabilistic behaviour had been demonstrated by Enrico Fermi and Adrian M Dirac, known as Fermi-Dirac statistics (for 'fermions', quantum particles of a half integer spin).

Against the coming back of the observer or the subject in the heart of the knowledge content, Einstein claimed objectivity in the description of a material world at the considered level (the 'quantum level'), a world that should exist 'really', independently of our means of knowing it.

EPILOGUE: PHYSICS AND RATIONALITY – THE REQUIREMENT OF OBJECTIVITY AND THE CONDITIONS FOR MATHEMATISATION

This conviction for the need of objectivity, posed as a condition of possibility for a meaningful science, that was deeply anchored in Einstein's thought and that guided and inspired him in his researches, is not without some resonance with the attitude of his far distant predecessor, Ibn al-Haytham, who formulated the requirement for a Physical Optics by stating the physical or natural character of light, independent of vision, that is to say, of the eye that comes to see through light as it already exists, in its autonomy and anteriority with respect to it. In one word, such a Physical Optics (which was a Geometrical Optics as well) was made possible by the removal of the 'visual ray'.

We cannot indeed exaggerate the scope of what appears as an affinity through the centuries. It is sufficient to note in both thinkers the requirement of *objectivity* that was compelling in both their respective approaches, each one in his original and fruitful way, and in his own time.

In the relationships between the use of Mathematics and of Physics, or more precisely in the *thought* of the one and of the other, it is striking to see how the first concern for Physics, that is to say, for the knowledge of *natural phenomena*, is determinant to decide the choice of the appropriate theorical and mathematical formulation. Under modalities as different as they may be with a nine centuries remoteness, both the researches here evoked of Ibn al-Haytham and of Einstein (and the same could be said about those of many scientists in the intermediate period that we could not discuss here) illustrate this direction of thought, this clarification (so to speak this *tuning*) of the conditions of the mathematisation of a given chapter of Physics, for us here that one which refers to light.

We see in it much more than a mere idealisation and 'application' of Mathematics to this domain of Physics: what is at stake is indeed a preparation and a construction of *concepts* for this domain so that they be able to allow the formulation of a theory that will be *mathematised* because the *physical concepts* thus formed have been legitimately expressed as *mathematical magnitudes*.

A clear and penetrating expression of this exigency can be found inside the time interval that separates our two evoked authors, proposed in the significant period of the beginning of 'modernity', when modern science was being shaped, decidedly under the double invocation of *Reason* (and, for physics specifically, mathematical rationality) and of *Nature* considered as it is, *objectively*, according to the (post-) Copernican decentration.

As I chose it among eventual other ones, for its immediate and powerful signification, particularly with respect to the subject we have treated here, it is a two-voices statement, emanating from two of the most creative and renowned scientists and thinkers, at a 100 years' distance from each other, René Descartes in the seventeenth century, and Jean d'Alembert in the eighteenth century, respectively: their statements (and their corresponding achievements), taken together, show intellectual harmony and complementarity, the second one adding a decisive improvement with respect to the first, namely the specific and detailed attention to the *physical* in the general meaning of that which is given first from the senses. This claim was indeed already present, at least from a qualitative point of view, among the scholars of Antiquity (especially in Aristotle), and after them particularly in Ibn al-Haytham, but always tempered by the suspicion of deformations through illusions (as the last one himself testified). The progress of observation and experimentation towards more precision, which were being acquired from the seventeenth century on, and later on the critical approaches from the philosophical point of view of John Locke, Etienne Bonnot de Condillac, David Hume and others, had ensured more firmly the validity of the knowledge as given in the data of observation duly submitted to critical reasoning (as in d'Alembert's own conception).

Descartes's considerations, in his *Rules for the Direction of the Mind* (*Règles pour la direction de l'esprit*),* on the conditions of the *intelligibility* of the objects of a physical science according to '*order and measure*', take the latter ('*measure*') as being the *relational aspect,* for example of algebraic type, between the (mathematical) magnitudes that characterise these objects (well further than the mere determination of a particular numerical value, which is a more recent restriction of the idea of 'measure' into 'measurement'). These relations express the *laws of nature* to which these objects are submitted. It is the establishment of such relations which provides *intelligibility* (in the light of reason): hence the strong link, in the physical sciences, between *mathematisation* (accompanying conceptualisation) and *intelligibility*.

Jean d'Alembert (a century later), Newtonian as for Physics, but also a follower of Descartes for what regards the function of Reason, enlightens somewhat this fundamental proposition on the meaning of the mathematical relationships by arguing and exemplifying it for the extension of the physical theory of Mechanics to a new domain. He performed actually one of the first fully conscious and explicit implementations of it, succeeding in handling in a theoretical way (by the means of magnitudes expressed mathematically, submitted to a few general physical principles) the problems of the motion of fluids which were previously treated empirically (using a number of specific assumptions).[†]

He succeeded thanks to a conceptual elaboration on a given physical problem (the motions of fluid bodies, made of material elements that move in all directions, differently from the solid bodies studied by the usual Mechanics[‡] that made use of the ordinary Differential Calculus proposed by Newton and Leibniz), and also

---

* Descartes (1628), mainly Rule 14. See Paty (2001b).

[†] D'Alembert (1752). See also D'Alembert (1758) 1765. On d'Alembert's thought and achievements, see in particular: Paty (1977, 1998, 2004).

[‡] Strictly speaking, the expression '*Analytical Mechanics*' has been used to name the science of Mechanics as it has been formulated since Lagrange's magistral work with this title. However, the works that precede it immediately and which it achieves, as those of d'Alembert and Euler, would deserve also such qualification, as they already deal also with the problems of Mechanics by using the Mathematical Analysis (which is the Differential and Integral Calculus enlarged to the partial differences).

thanks to the joint formulation of a new appropriate mathematical theory, that he was also establishing, namely the Theory of Partial Differential Equations. He thereby founded Fluid Mechanics as a rational, mathematised, physical theory of a domain of Physics that was hitherto empirical (it would fall to Leonard Euler to achieve its formulation). Theorising the various domains of Physics beyond *Mechanics*, called *Analytic Mechanics* after the synthesis formulated by Joseph Louis Lagrange in his book of that title, would be the work of theoretical and mathematical physicists of the nineteenth century, and it would also be performed under the sign of the Theory of Partial Differential Equations, which is actually the Mathematical Theory of continuous magnitudes.

Let us try a last word to conclude. It would also be worthwhile to stress the importance, in the exercise of thought, throughout the journey from one stage to the other in the historical elaboration of scientific knowledge (here, about *light* conceived in its *natural or physical character*, let us say in its *materiality*), of the close interaction between, on the one hand, the *formulation* and the *solving* of *scientific problems* (*physical ones*, in this case) and, on the other hand, the implementation of an appropriate rationality, which also includes the dimension of *creation*, that is to say the development of new forms gained on the unknown. This dimension appears as soon as one enters into the texts these scholars have left us. It appears with evidence in those we evoked in this contribution.

REFERENCES

Bachelard, Gaston, 1938. *La Formation de l'esprit scientifique*; ré-éd., Vrin, Paris, 1972.

Bachelard, Gaston, 1949. *Le Rationalisme appliqué*, Presses Universitaires de France, Paris, 1949; 1970.

D'Alembert, Jean, 1743, 1758. *Traité de dynamique*, David, Paris, 1743; 2e éd., modif. et augm., David, Paris, 1758; repr., Culture et Civilisation, Bruxelles, 1966.

D'Alembert, Jean, 1752, 1966. *Essai d'une nouvelle théorie de la résistance des fluides*, David, Paris; repr., Culture et Civilisation, Bruxelles.

D'Alembert, Jean, 1758, 1765. *Essai sur les éléments de philosophie ou sur les principes des connaissances humaines*, Paris, 1758; new ed. augm. by the *Eclaircissements* (1765) (R N Schwab (ed.)), Olms Verlag, Hildesheim, 1965.

Darrigol, Olivier, 2000. *Electrodynamics from Ampère to Einstein*, Oxford University Press, Oxford (UK).

Descartes, René, 1628. Regulæ ad directionem ingenii, in R Descartes (ed.), *Oeuvres de Descartes*, publiées par Ch. Adam et P Tannery, 11 volumes (1st ed., 1896–1913); new éd. révis., 1964–1974; new ed., 1996, vol. 10, pp. 348–349; French. transl., *Règles pour la direction de l'esprit*, Paris, Vrin, 1970.

Einstein, Albert, 1905. Elektrodynamik bewegter Körper, *Annalen der Physik*, ser. 4, XVII, 1905, 891–921; in CP, vol. 2, pp. 275–306.

Einstein, Albert, 1924. Quantentheorie des einatomigen idealen Gases, Preussische Akademie Wissenschaften, *Phys. Math. Klasse, Sitzungsberichte* 22, 261–267.

Einstein, Albert, 1987–2013. *The Collected Papers of Albert Einstein* (J Stachel, M J Klein and, D Kormos Buchwald et al. (éds.)), Princeton University Press, Princeton, New Jersey, 1987–2013, vols. 1–13 (and the corresponding volumes of English transl.). (Indicated as CP). Cf. in particular vols. 2, 3, 4, 6, and 13 (writings up to 1923).

Jammer, Max, 1966. *The Conceptual Development of Quantum Mechanics*, Mc Graw-Hill, New York.

Lindberg, David C, 1997. La réception occidentale de l'Optique arabe, in R Rashed and R Morelon (éds.), Histoire des sciences arabes, Seuil, Paris, vol. 2, pp. 355–367.

Morélon, Régis, 1999. *Astronomie physique et astronomie mathématique dans l'astronomie précopernicienne*, in R Rashed and J Biard (éds.), pp. 105–130.

Neugebauer, Otto, 1957. *The Exact sciences in Antiquity*, Brown University Press; 2nd ed., Dover, New York, 1969.

Newton, Isaac, 1687, 1726. *Philosophiae Naturalis Principia Mathematica*, London, 1687; 2nd ed., 1713; 3rd ed., 1726; A Koyré and I B Cohen (éds.), Cambridge University Press, Cambridge, 1972.

Pais, Abraham, 1979. Einstein and the quantum theory, *Review of Modern Physics* 51, 861–914.

Paty, Michel, 1977. *Théorie et pratique de la connaissance chez Jean d'Alembert*, Thèse de doctorat en philosophie, Université des Sciences Humaines, Univ. Strasbourg 2.

Paty, Michel, 1985. La tradition retrouvée des algébristes arabes, *La Recherche* (Paris), 16, (n° 167, juin), 820–821 (suivi d'une correspondance avec P Costabel, ibid. (n° 169, septembre), 1103–1104.

Paty, Michel, 1987. La tradition mathématique arabe, *Archives de Philosophie* (Paris), 50, (cahier 2, avril-juin), 199–217. Repr. in Paty (1990a), Chapter 6, pp. 87–104.

Paty, Michel, 1990. *L'analyse critique des sciences, ou le tétraèdre épistémologique (sciences, philosophie, épistémologie, histoire des sciences)*, L'Harmattan, Paris.

Paty, Michel, 1993. *Einstein philosophe. La physique comme pratique philosophique*, Presses Universitaires de France, Paris.

Paty, Michel, 1994. Sur l'histoire du problème du temps : le temps physique et les phénomènes, in E Klein and M Spiro (éds.), *Le temps et sa flèche*, Editions Frontières, Gif-sur-Yvette, 1994, pp. 21–58; repr. : Collection 'Champs', Flammarion, Paris, 1996, pp. 21–58.

Paty, Michel, 1996. Poincaré et le principe de relativité, in J-L Greffe, G Heinzmann and K Lorenz (éds.), *Henri Poincaré. Science et philosophie. Science and philosophy. Wissenschaft und Philosophie. Congrès international, Nancy, France, 14–18 mai 1994*, Akademie Verlag, Berlin/Albert Blanchard, Paris, pp. 101–143.

Paty, Michel, 1998. *D'Alembert ou la raison physico-mathématique au siècle des Lumières*, Collection 'Figures du savoir', Les Belles Lettres, Paris; 2nd print, 2004.

Paty, Michel, 1999a. Universality of Science : Historical Validation of a Philosophical Idea, Chapter 12, in S I Habib and D Raina (eds.), *Situating the History of Science: Dialogues with Joseph Needham*, Oxford University Press, New Delhi, 1999, pp. 303–324; Oxford India Paperbacks, 2001, 303–324. – (Original in French: L'Universalité de la science. Une idée philosophique à l'épreuve de l'histoire, *Mâat. Revue Philosophique Africaine* (Dianoia, Université de Douala, Cameroun), 1st year, n° 1, April 1999, 1–26).

Paty, Michel, 1999b. Comparative history of modern science and the context of dependency, *Science, Technology and Society. An International Journal Devoted to the Developping World* Sage Publications, New Delhi, vol. 4, 2 (July–December), 171–204. Transl. from French by N Flay, verified by the author. – Original in French: L'Histoire comparative des sciences modernes et le contexte de dépendance, *Convergencia. Revista de Ciencias Sociales* (México), Año 8, núm. 24, enero-abril 2001, 11–48.

Paty, Michel, 2001a. Intelligibilité et historicité (Science, rationalité, histoire), in J J Saldaña (ed.), *Science and Cultural Diversity. Filling a Gap in the History of Science*, Cadernos de Quipu 5, México, pp. 59–95.

Paty, Michel, 2001b. La Notion de grandeur et la légitimité de la mathématisation en physique, in M Espinoza (éd.), *De la science à la philosophie. Hommage à Jean Largeault*, L'Harmattan, Paris, pp. 247–286. – English transl. : The Idea of quantity at the origin of the legitimacy of mathematization in physics, *in* C Gould (ed.), *Constructivism and Practice: Towards a Historical Epistemology*, Rowman & Littlefield, Lanham (Md.,USA), 2003, pp. 109–135.

Paty, Michel, 2001c. Réflexions sur le concept de temps, *Revista de Filosofia (Madrid)*, 3a época, volumen xiv, n 25, 53–92.

Paty, Michel, 2002. Poincaré, Langevin et Einstein, *Épistémologiques. Philosophie, sciences, histoire. Philosophy, science, history* (Paris, São Paulo) 2, n°1–2, janvier-juin 2002, 33–73 (Special issue: B Bensaude-Vincent;, M-C; Bustamante;, O Freire and M Paty (éds.), *Paul Langevin, son œuvre et sa pensée. Science et engagement*).

Paty, Michel, 2004. L'Élément différentiel de temps et la causalité physique dans la dynamique de Alembert, in R Morelon and A Hasnawi (éds.), *De Zénon d'Elée à Poincaré. Recueil d'études en hommage à Roshdi Rashed*, Editions Peeters, Louvain (Be), pp. 391–426.

Paty, Michel, 2007. Rationalités comparées des contenus mathématiques. Sur les travaux de Roshdi Rashed, ou : La Philosophie dans le champ de l'histoire des sciences, *Dogma. Revue des revues. Epistémologie* (Revue électronique, Paris), 36 p. – (Contribution au *Colloque des Sciences Arabes, Damas (Syrie), 1–4 novembre 2002.*)

Paty, Michel, forthcoming. *Einstein, les quanta et le réel. Critique et construction théorique.*

Poincaré, Henri, 1989–1891. *Cours de physique mathématique. Leçons sur la théorie mathématique de la lumière*, Carré et Naud, Paris, vol. I, 1889; vol. II, 1891.

Poincaré, Henri, 1890–1891, 1901. *Electricité et optique, I. Les théories de Maxwell et la théorie électromagnétique de la lumière*, Carré, Paris, 1890; II. *Les théories de Helmholtz et les expériences de Hertz*, Carré, Paris, 1891. – *Electricité et optique. La lumière et les théories électrodynamiques*, Deuxième édition, revue et complétée, Carré et Naud, Paris, Gauthier-Villars, Paris, 1901.

Rashed, Roshdi, 1970. Optique géométrique et doctrine optique chez Ibn al Haytham (1970), re-edited. in R Rashed (ed.), *Optique et mathématiques. Recherches sur l'histoire de la pensée scientifique en arabe*, Variorum, Ashgate Publ., Aldershot (GB), 1992, Chapter 2, pp. 271–298.

Rashed, Roshdi, 1978. Lumière et vision: l'application des mathématiques dans l'optique d'Ibn al-Haytham, 1978, republ. dans R Rashed, *Optique et mathématiques. Recherches sur l'histoire de la pensée scientifique en arabe*, Variorum, Ashgate Publ., Aldershot (GB), 1992, Chapter 4, pp. 19–44.

Rashed, Roshdi, 1989. Problems of the transmission of Greek scientific thought into Arabic: Examples from mathematics and optics. *History of Science* 27, 199–209.

Rashed, Roshdi (ed.), 1991. *Mathématique et philosophie, de l'Antiquité à l'âge classique, Hommage à Jules Vuillemin*, Editions du C.N.R.S., Paris.

Rashed, Roshdi, 1992. *Optique et mathématiques. Recherches sur l'histoire de la pensée scientifique en arabe*, Variorum, Ashgate Publ., Aldershot (UK).

Rashed, Roshdi, 1993. *Géométrique et dioptrique au xe siècle. Ibn Sahl, al-Quhi et Ibn al-Haytham*, Les Belles Lettres, Paris.

Rashed, Roshdi, 1997. L'Optique géométrique, in R Rashed and R Morelon (éds.), *Histoire des sciences arabes*, Seuil, Paris, vol. 2, pp. 293–318.

Rashed, Roshdi and Biard, Joël (eds.), 1999. *Les Doctrines de la science de l'Antiquité à l'âge classique*, Peeters, Leeuven (B).

Rashed, Roshdi and Morelon, Régis (eds.), 1997. *Histoire des sciences arabes*, Seuil, Paris, 3 vols.

Rosenfeld, Boris A and Youskevitch, Adolf P, 1997. Géométrie, in R Rashed and R Morelon (eds.), *Histoire des sciences arabes*, Seuil, Paris, vol. 2, pp. 121–162.

Russell, 1997. La naissance de l'Optique physiologique, in R Rashed and R Morelon (eds.), *Histoire des sciences arabes*, Seuil, Paris, vol. 2, pp. 319–354.

Sabra, Abdelhamid I, 1972. Ibn al-Haytham, in C C Gillispie (ed.) *Dictionary of Scientific Biography*, Scribner's Sons, New York, vol. 6, pp. 189–210.

Simon, Gérard, 1988. *Le Regard, l'Être et l'Apparence*, Seuil, Paris.

Vuillemin, Jules, 1999. D'Eudoxe à Képler, in R Rashed and J Biard (eds.), *Optique et mathématiques. Recherches sur l'histoire de la pensée scientifique en arabe*, Variorum, Ashgate Publ., Aldershot (GB), pp. 87–104.

Youschkevitch, Adolf P, 1976. *Les mathématiques arabes (xviiie–xve siècles)*, trad. fr. de M Cazenave et K Jaouiche, Vrin, Paris (Original in Russian, 1961).

4 Translating and Interpreting Ibn al-Haytham's *Optics* from Arabic to Latin

New Light on the Vocabulary of Reflection and Refraction

Paul Pietquin

CONTENTS

Overview of the Latin Translation 44
Vocabulary of 'Reflection' and 'Refraction' 45
Reflexio Stands for Both Meanings 45
Use of *Conversio* and *Obliquatio* 47
'Ambiguity in the Minds of Those Using the Terms'? 49
Conclusion 50
References 51

When, in 1572, Friedrich Risner published the very first printed edition* of the Latin translation of Ibn al-Haytham's *Book of Optics* (*Kitāb al-Manāẓir*), he chose to publish it together with another treatise, written in Europe, but largely based on Ibn al-Haytham's work: Witelo's *Perpectiva*, composed around 1270.[†] Risner titled this collection of works *Opticae thesaurus* ('Treasure of Optics'), a real treasure indeed, as it remained, for more than 400 years, the only printed edition available of *Kitāb al-Manāẓir* (translated into Latin as *De aspectibus*, with the name of the author rendered as Alhacen or Alhazen[‡]). It was not until recently that critical editions of the text were published in Arabic (by Sabra[§]) and in Latin (by Smith and by me[¶]), alongside English and French translations.

---

* Risner 1572.
† See Kelso 1994; Unguru 1977, 1991; Smith 1983.
‡ Smith 2001, xxi.
§ Sabra 1983, 1989, 2002.
¶ Smith 2001, 2006, 2008, 2010; Pietquin 2010.

43

OVERVIEW OF THE LATIN TRANSLATION

The Latin translation, written in Mozarabic Spain at the very beginning of the thirteenth century or somewhat earlier, is relatively faithful to the Arabic original, although the first three chapters of book 1 are missing.* According to Smith,[†] abrupt changes in style and vocabulary at some points in book 3 and book 6 suggest that at least two – but probably three – translators were involved, whose names remain unknown.

Even a generally accurate translation can be misleading by just missing one word or even one letter. For instance, one of the laws concerning refraction of light from air to water,[‡] given in Book 7, Chapter 3, states that the deviation of the angle of refraction is always smaller than the angle of incidence, mistranslating the Arabic original version that said it was less than *half* of the angle of incidence.[§]

Furthermore, the Latin translation has circulated in the form of manuscript codices (of which 23 still exist – as compared to only five witnesses of the Arabic work[¶]), so that none of them is an exact copy of the other.

Here is an example. Figure 4.1 reproduces – from page 277 of Risner's edition – a geometrical demonstration of how an object (A) is seen through a sphere of glass. In his accompanying text, Risner, as in all manuscripts but one, incorrectly claims that the perpendicular at the point of refraction G is EH, instead of EG (as can be inferred from the drawing).

> [...] *tunc forma, quæ extenditur per ag: refringiur per gm in partem perpendicularis,* que est eh: & *cum forma perueniet ad m: refringetur secundo in partem contrariam perpendicularis, quæ est emc* [...] (ed. Risner 1572, 277, no. 49).

Smith's edition of 2010** gives the correct letters EG, as found in only one Latin manuscript from Cambridge Trinity College and also in the Arabic version.[††]

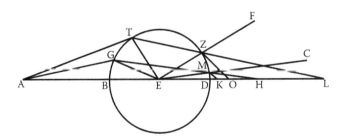

FIGURE 4.1   Refraction of an object through a sphere of glass.

---

* Sabra 1989, t. II, lxxiv; Smith 2001, ix and xxiii–xxiv.
† Smith 2010, cxxv–cxxvi.
‡ More generally, from any transparent body to another body that is denser than the first.
§ *De aspectibus*, Book 7 [3.34.4] Smith; Cf. Smith 2010, 361, n. 70; Pietquin 2010, 138, n. 29.
¶ Smith 2015, 16 and n. 39. Pietquin 2010, 17–19.
** Smith 2010, 128, l. 279.
†† Rashed 2005, 218–219; Pietquin 2010, 300, n. 16.

[…] the form that extends along AG is refracted along GM toward normal EG, and when the form reaches M, it will be refracted again away from normal EMC (trans. Smith 2010, Book 7 [7.37]).

The same error that we noticed in Risner occurs in all the other Latin manuscripts, including the recently uncovered Biblioteca Casanatense, Ms 1393.* This error, among other clues, could be an indication that the Trinity College manuscript may be the closest witness[†] of the Arabic model used for translation, as well as proving that Risner did not use it.

At the time when Risner published his edition, Latin was still a vehicular language for written scholarship. Therefore, it is not surprising that he did not make an exact copy of the two manuscripts he had found,[‡] but rather polished the text in several ways. Specifically, while all extant Latin manuscripts use the same unique word (*reflexio*) to render the two totally distinct concepts of 'reflection' and 'refraction', Risner used two different Latin terms: *reflexio* and *refractio* (or words of the same families, respectively: *reflectere*/*refringere*).[§] In this paper, I will precisely focus on the vocabulary used in the Latin translation to denote 'reflection' and 'refraction', because these are key concepts to evaluate how Ibn al-Haytham's revolutionary science was understood when introduced to the West.

VOCABULARY OF 'REFLECTION' AND 'REFRACTION'

Before Ibn al-Haytham, mathematical optics were driven by the idea that visual rays came from the eye to reach the objects seen,[¶] so there was probably no great difference between considering that these rays were breaking (*refractio*) or bending (*reflexio*) at the surface of mirrors before reaching the objects. Similarly, the breaking of the visual ray when passing from the air to water could easily be considered as a peculiar form of bending.

REFLEXIO STANDS FOR BOTH MEANINGS

This could explain why the same word *reflexio* is used in the Latin translation to render either انعكاس/*in'ikās*, ('reflection') or انعطاف/*in'iṭāf* ('refraction') from Arabic,** even when both occur in the sae sentence, as shown in this example: one instance of *in'iṭāf* and two instances of *in'ikās* are rendered three times by the same word in Latin (*reflexio*/*reflexionis*).

If straight line *BG* is parallel to straight line *EA*, then the image is not delimited, and the eye perceives the form at the point of refraction (الانعطاف). The cause of this is

---

* Available online: http://db.biblhertu.it/alhuven/index.xml.

† Pietquin 2010, 35.

‡ Smith 2001, xxiii.

§ See Table 4.1.

¶ Lejeune 1948, 18ff.

** See Latin-English word indexes in Smith 2001, 2006, 2008, 2010 s.v. *reflexio* and *reflectere*. Latin-Arabic glossary in Sabra 1989, t. II, 173–202, is misleading, as it is based on Risner's edition featuring *refractio* and *refringere*.

similar to the cause we mentioned in the case of reflection (الانعكاس), when reflection (الانعكاس) is on a straight line parallel to the perpendicular (trad. and ed. Rashed 2005, 194).

> *Et si linea BZ fuerit equidistans linee EA, tunc ymago erit indeterminata, et forma comprehendetur in loco* reflexionis. *Huius autem causa similis est illi quam diximus in loco* reflexionis, *cum fuerit* reflexio *per lineam equidistantem perpendiculari* (ed. Smith 2010, Book 7 [5.63]).

> *If line BZ is parallel to line EA, the image will be indefinite, and the form will be perceived at the point of* refraction. *The reason for this is similar to the one we discussed in regard to* reflection *when* reflection *occurs along a line parallel to the normal* (trad. Smith 2010, Book 7 [5.63]).

> *Si linea bz fuerit æquidistans lineæ ea: tunc imago erit indeterminata, & forma comprehendetur in loco* refractionis. *Huius autem caussa similis est illi quam diximus in loco* reflexionis [...] *cum fuerit* reflexio *per lineam æquidistantem perpendicular* (ed. Risner 1572, 261).

Somewhat paradoxically, when Risner corrected the text of the manuscript and introduced one instance of *refractio* instead of *reflexio*, he ended up with a text closer to the Arabic original.

At this point, two questions may arise: were the translators aware of the difference in meaning between *in'ikās* and *in'iṭāf*? and, if so, did all the readers of the translation understand this difference as well as Risner did a long time afterwards?

To address the second question, we may have a look at the works of the so-called 'Perspectivists' who, a few decades after the Latin translation was completed, wrote new books on optics that undoubtedly relied, directly or indirectly, on it: Witelo, Roger Bacon and John Pecham* (see Table 4.1).

In my opinion, Roger Bacon and John Pecham employ the most consistent terminology: *reflexio* for 'reflectio' and *fractio* for 'refraction'.[†] When a light ray hits a mirror, it bounces back at the same angle it formed with the perpendicular to the mirror in the first place. This can be best described as a bending going away from the mirror ('to bend' is *flectere* in Latin – hence the noun *flexio* – and *re-* is a prefix which means 'away' or 'backwards'). On the contrary, when the light passes from air to water (or from water to air), it does not bounce back but just changes its direction, and so there is no need to add the *re-* prefix to *fractio* (from Latin *frangere* i.e., 'to break').[‡]

---

* Smith 2015, 16 and n. 40. Table 4.1 compiles the results of different searches on electronic digitised versions of the following texts: Unguru 1991, Lindberg 1996, Lindberg 1970.

† Or *reflectere* for 'to reflect' and *frangere* for 'to refract'. Cf., for example, Lindberg 1996, 286, l. 5–6: *Manifestato quomodo visio fiat per lineas* rectas *et* reflexas, *nunc tertio, manifestandum est quomodo fiet per* fractas. In Lindberg 1970, *reflexio* is mostly used in Parts I and II, while *fractio* occurs more frequently in Part III.

‡ In one unique example (Lindberg 1970, 88, l. 378), Pecham uses *flectere* (but not *reflectere*) as a synonym of *frangere* to denote refraction, both terms being used precisely without the *re-* prefix: *Quando vero obviat magis vel minus dyaphano recedit a rectitudine, et quasi frangitur vel flectitur in obliquum,* 'When light and color are more or less opposed by a transparent body, they depart from their rectilinear paths and are, as it were, broken or bent sideways.'

TABLE 4.1
Latin Vocabulary of 'Reflection' and 'Refraction' among Perspectivists

Kitāb al-Manāẓir, ca. 1030	انعكاس (*in'ikās*)	انعطاف (*in'iṭāf*)
De aspectibus, ca. 1200	*Reflexio*	*Reflexio*
Thesaurus opticae, 1572	*Reflexio*	*Refractio*
Witelo, ca. 1230 – after 1280	*Reflexio*	*Refractio*
Roger Bacon, ca. 1214–1292	*Reflexio*	*Fractio*
John Pecham, ca.1230–1292	*Reflexio*	*Fractio*

Although Witelo was influenced by Bacon,* he used the prefix in both cases.† On the other hand, Witelo's use is similar to that of Risner in *Opticae thesaurus*. One can wonder, therefore, if the reason for Risner changing *reflexio* into *refractio* (to denote precisely what we call now 'refraction') was not simply his desire to use the same terminology as in Witelo's *Perpectiva* that came together in the same volume.

USE OF *CONVERSIO* AND *OBLIQUATIO*

A closer look at the vocabulary of reflection and refraction in the manuscripts of the Latin translation reveals that they also use two other words that are almost never used by Perspectivists: *conversio* and *obliquatio*.

Conversio is used to denote 'reflection' (*in'ikās* in Arabic), and seems especially useful whenever there is a need to specify 'reflection' as opposed to 'refraction', which is then still rendered in Latin as *reflexio* in the example below.

> […] however, it so happens that from this position the eye perceives the form of the visible object in the position of the refraction (انعطاف), for the same cause that we mentioned for reflection (انعكاس) in mirrors, if the reflection (انعكاس) is from the circumference of a circle in a sphere, and if the image is the center of the eye (ed. Rashed 2005, 200).
>
> *Sed ex hac positione accidit quod visus comprehendat formam rei vise apud locum* reflexionis, *ea de causa quam diximus in* conversione *ex speculis, cum fuerit* conversio *a circumferentia in aliqua spera, et fuerit ymago centrum visus* (ed. Smith 2010, Book 7 [5.71]).
>
> *Sed ex hac positione accidit, ut uisus comprehendat formam rei uisae apud locum* refractionis *ea de caussa, quam diximus in* reflexione *ex speculis, cum fuerit* reflexio *a circumferentia in aliqua sphaera, & fuerit imago centrum uisus* (ed. Risner 1572, 263, n° 29).

Risner, of course, to remain consistent, changed *conversio* into *reflexio* in his edition.

Obliquatio is used the same way to denote 'refraction' (*in'iṭāf* in Arabic), as opposed to 'reflection'. For example, Book 7 of Ibn al-Haytham's Optics, devoted to

---

* Cf. Lindberg 1971.
† Cf. Unguru 1991, 291, l. 4–6: *Volentes autem formarum naturalium actiones sub triplici videndi modo prosequi, scilicet illo qui fit per* simplicem visionem, *et eo qui per* reflexionem, *et illo qui per* refractionem, […].

TABLE 4.2
Latin Vocabulary of 'Reflection' and 'Refraction' around 1200

Kitāb al-Manāẓir, ca. 1030	انعكاس (*in'ikās*)	انعطاف (*in'iṭāf*)
De aspectibus, ca. 1200	*Reflexio*	*Reflexio*
	Conversio	*Obliquatio*
Gerard of Cremona, d. 1187	*Conversio*	
Eugene of Sicily, d. 1203	*Refractio/reverberatio*	*(Re)fractio/flexio*
William of Moerbeke, ca. 1269	*Refractio*	

the study of refraction, is indifferently referred to as *sermo de obliquatione* or *sermo de reflexione*.[*]

Since almost none of the Perspectivists employ either *conversio* or *obliquatio*, we will have a look now at other Latin translations of books on optics that circulated at about the same time (see Table 4.2).

In this sample of books, *conversio* is only to be found in a translation from a lost Arabic work of al-Kindī made by Gerard of Cremona,[†] who never uses the term *reflexio*. The book deals with reflection of rays in mirrors, but not with refraction of rays through water. Gerard of Cremona is sometimes believed to be one of the translators of Ibn al-Haytham, but this is far from certain.[‡]

The famous translation of Ptolemy's Book of Optics made by Eugene of Sicily from a lost Arabic model features a more complex vocabulary set.[§] At the beginning of Chapter 5, devoted to refraction, he says that *fractio* can occur either by means of *reverberatio* (i.e., 'reflection') or by means of *flexio* (i.e., 'refraction').[¶] So here, the common term for both concepts is *fractio* and not *reflexio*. *Flexio* is only used to denote refraction. Note also that the Latin prefix *re-*, meaning 'away' or 'backwards', is nearly exclusively used for reflection of rays in a mirror.

In a short treatise *De speculis* ('About mirrors') translating a lost Greek original, William of Moerbeke[**] describes the rebound of rays in a mirror as *refractio*, which is congruous with the vocabulary set of Eugene of Sicily.

Although each author or translator, taken individually, seems so far to have been quite consistent in his use of terms, the fact that both *reflexio* and *refractio* have been used either to denote what we call now 'reflection' or 'refraction' may have some- times been confusing to them.

---

[*] Smith 2001, 27, 1. 166; 29, 1. 221; 38, 1. 197. *Obliquatio* is also sometimes used by Witelo, cf. Unguru 1991, 319, 1. 70: *illam refractionem aut obliquationem.*

[†] Cf. Rashed 1997, 454, n. 14; 508, n. 45; 759–760.

[‡] Smith 2001, xx.

[§] Lejeune 1989, 273ff. (*index verborum*).

[¶] Lejeune 1989, 223: *Cum de eo quod accidit in* fractionibus *uisibilis radii, aliud quidem est secundum aduersationem et fit ex* reuerberatione *existente a rebus que prohibent penetrationem et continentur sub nomine speculorum, aliud autem existit secundum penetrationem et fit ex* flexione *existente in rebus que non prohibent penetrationem et subiacent uni nomini, uidelicet quod penetrat uisus …*

[**] Jones 2001.

'AMBIGUITY IN THE MINDS OF THOSE USING THE TERMS'?

This was the opinion of Eastwood:

> Until the mid-XIIIth century no fixed terminology for reflection and refraction was current in northern Europe. Sometimes the word *fractio* served for both phenomena alternately. [...] This plurality of usage seems to reflect an ambiguity in the minds of those using the terms. If one uses the same term for reflection or refraction, does one really consider the two phenomena radically different?*

This may have been the case for Grosseteste, whom Eastwood had certainly in mind while formulating this remark. However, as for the translators of *Kitāb al-Manāẓir*, we have at least one indication that speaks against 'ambiguity in their minds'.

In the following passage from *De perspectiva*, the translator adds an explanation to *reflexionem*, a word that could – as already discussed – stand for both 'reflection' and 'refraction':

> *Et cum diversitas diafonitatis affirmat* reflexionem, scilicet obliquationem, *et diversitas qualitatis sensus affirmat illam obliquationem* [...] (ed. Smith 2001, Book 2 [2.14]).
> *But since the difference in transparency prompts refraction, i.e., bending, and since the difference in sensitivity prompts [such] bending* [...] (trad. Smith, 2001).

The translation of Smith is of course here inadequate. I would rather render *reflexionem, scilicet obliquationem* by 'bending, that is to say: refraction'.

This explanation, as far as I could check,† is not in the Arabic original, so it seems actually due to the translator, who had in this case absolutely no ambiguity in mind.

As for the translators of other Arabic works, they could have been confused by the graphic proximity of the two words in Arabic and also by inconsistent use of these words before and after Ibn al-Haytham. I gathered the information in Table 4.3 from Kheirandish's edition of *The Arabic Version of Euclid's Optics*.‡

TABLE 4.3

Overview of Arabic Vocabulary of 'Reflection' and 'Refraction'

	'Reflection'	'Refraction'
Kitāb Uqlīdis ..., ca. 828	انعطاف (*in'iṭāf*)	
Ḥunayn ibn Isḥāq, d. 877	انعكاس (*in'ikās*)	
Qusṭā ibn Lūqā, d. 912	انعكاس (*in'ikās*) / انعطاف (*in'iṭāf*)	
Al-Fārābī, d. 950	انعطاف (*in'iṭāf*)	
Ibn al-Haytham, d. 1040	انعكاس (*in'ikās*)	انعطاف (*in'iṭāf*)
Naṣīr al-Dīn al-Tūsī, d. 1274	انعكاس (*in'ikās*) / انعطاف (*in'iṭāf*)	انعطاف (*in'iṭāf*)

---

* Eastwood 1967, 406.
† Cf. English translation of Sabra 1989, t. I, 118: 'But if the difference in transparency requires that the form should be refracted, and the difference in the manner of sensation requires that it suffers that [same] refraction [...]'.
‡ Kheirandish 1999, t. II, 55–59; see also Rashed 1997, 588–589 and 736.

On the whole, it confirms the opinion of Rashed, stating that 'the two terms [انعكاس (in'ikās)/انعطاف (in'iṭāf)] were indistinctly used to denote reflection of rays [...] before the latter was restricted to refraction'.* We have here something similar to the word use of Eugene of Sicily, restricting *flexio* to 'refraction', although it had been also used elsewhere for 'reflection'.

In any case, it is clear that the Latin word used in the field of optics was driven both by Arabic and Greek.† However, unfortunately, such translations from Greek, where the Greek original is still available, are quite rare, at least on the subject of the refraction of light.

We shall finally turn again to Perspectivist Roger Bacon, who also gives a clear indication that there was absolutely no 'ambiguity' in his mind on the subject of reflection and refraction.

In the text below, Bacon quotes a passage from Alhacen:

[...] *Ex quo patet quod visus comprehendit stellas* reflexe (id est fracte), *non recte* [...] *Hec sunt verba Alhacen in septimo.*‡

The word *reflexe* stands here for 'by means of refraction' and not 'by means of reflection' as it should, given Bacon's consistent word use elsewhere. To avoid any ambiguity, Bacon adds the paraphrase *id est fracte* ('that is to say by refraction'). By doing this, he shows clearly that he understands Alhacen correctly and that he prefers another word to express the same concept.§

CONCLUSION

Transmission of knowledge from one culture to another – or from one age to another – is of course not only a question of words and vocabulary. As for Ibn al-Haytham's *Kitāb al-Manāẓir*, featuring three books devoted to direct vision, followed by three other books devoted to reflection and a last one devoted to refraction, the structure of the work itself – as well as consistent use of words in Arabic – contributed to avoiding ambiguity.

However, when the translators started to turn the *Book of Optics* from Arabic to Latin, there was no accepted terminology to render the concepts of reflection and refraction. They managed to avoid ambiguity by developing a relatively consistent

---

* Rashed 2000, 158.

† I will only briefly discuss the case of the Arabic translation of Euclid's Optics (*Kitāb Uqlīdis...*), where the word *in'iṭāf* (instead of the more 'standard' *in'ikās*) is used for a ray reflecting on a mirror (Proposition 20, see Kheirandish 1999, t. II, 55). In the Greek version, only the verb ἀντανακλάω/*antanaklaô* (or ἀνακλάω/*anaklaô*) is used, with κλάω/*klaô* meaning 'to break' or 'to deflect', ἀντι-/*anti*-meaning 'against' and ἀνα-/*ana*-meaning 'upwards' or 'backwards'. In such a context, the corresponding verb used in Arabic (*in'aṭafa*) does not seem to me so incongruous.

‡ Lindberg 1983, 124, l. 92–93 and 106–107; cf. *De aspectibus*, Book 7 [4.28] Smith.

§ Bacon behaves similarly when quoting the Latin translation of Ptolemy (Lindberg 1983, 120, l. 30–32): *Dicit igitur Ptolomeus: 'Possibile est nobis dinoscere quod in loco contiguationis aeris ad etherem fit flexio [reflexio FO] radii, id est, fractio [...]'* The word *(re)flexio* is used here (by the translator of Ptolemy) to denote refraction and not reflection. And here again we have a paraphrase by Bacon (*id est, fractio*). Cf. Smith 1996, V [23].

use of terms. Since no clear choice in Latin terminology had arisen among scientists in the thirteenth century, Risner's printed edition probably contributed to spreading the use of two different words, which are similar to those that are still employed today in Western languages.

REFERENCES

Eastwood, B S 1967. Grosseteste's 'Quantitative' Law of Refraction. A chapter in the history of non-experimental science. *Journal of the History of Ideas* 28: 404–414.

Jones, A 2001. Pseudo-Ptolemy *De Speculis*. *SCIAMVS* 2: 145–186.

Kelso, C, ed. and trans. 1994. Witelonis Perspectivae liber quartus. PhD diss., University of Missouri, Columbia.

Kheirandish, E 1999. *The Arabic Version of Euclid's Optics*. New York: Springer. Vol. I: 246 pp.; Vol. II: 165 pp.

Lejeune, A 1948. *Euclide et Ptolémée. Deux stades de l'optique géométrique grecque. Université de Louvain, Recueil de Travaux d'Histoire et de Philologie, 3ᵉ série, 31.* Louvain: Bibliothèque de l'Université, 196 pp.

Lejeune, A, ed. and trans. 1989. *L'*Optique *de Claude Ptolémée dans la version latine d'après l'arabe de l'émir Eugène de Sicile*. Leiden: Brill, 371 pp.

Lindberg, D C, ed. and trans. 1970. *John Pecham and the Science of Optics*. Madison: University of Wisconsin Press, 300 pp.

Lindberg, D C 1971. Lines of influence in thirteenth-century optics: Bacon, Witelo, and Pecham. *Speculum* 46: 66–83.

Lindberg, D C, ed. and trans. 1983. *Roger Bacon's Philosophy of Nature: A Critical Edition, with English Translation, Introduction, and Notes, of* De multiplicatione specierum *and* De speculis comburentibus. Oxford: Clarendon Press, 420 pp.

Lindberg, D C, ed. and trans. 1996. *Roger Bacon and the Origins of* Perspectiva *in the Middle Ages: A Critical Edition and English Translation of Bacon's* Perspectiva *with Introduction and Notes*. Oxford: Clarendon Press, 411 pp.

Pietquin, P, ed. and trans. 2010. *Le septième livre du traité* De aspectibus *d'Alhazen, traduction latine médiévale de l'*Optique *d'Ibn al-Haytham*. Brussels: Académie royale de Belgique, 368 pp.

Rashed, R 1997. *Œuvres philosophiques et scientifiques d'al-Kindī*. Vol. 1. Leiden: Brill, 776 pp.

Rashed, R 2000. *Les catoptriciens grecs I: Les miroirs ardents*. Paris: Les Belles Lettres, 450 pp.

Rashed, R 2005. *Geometry and Dioptrics in Classical Islam*. London: al-Furqan, 1178 pp.

Risner, F, ed. 1572. *Opticae Thesaurus*. Basel: per Episcopios, 474 pp.

Sabra, A, ed. 1983. *The Optics of Ibn al-Haytham. Books I–II–III: On Direct Vision*. Kuwait: National Council for Culture, Arts, and Letters, 781 pp.

Sabra, A, trans. 1989. *The Optics of Ibn al-Haytham: Books I–III on Direct Vision*. London: Warburg Institute. Vol. I: 367 pp.; Vol. II: 246 pp.

Sabra, A, ed. 2002. *The Optics of Ibn al-Haytham. An Edition of the Arabic Text of Books IV–V: On Reflection and Images Seen by Reflection*. Kuwait: National Council for Culture, Arts, and Letters, 760 pp.

Smith, A M, ed. and trans. 1983. *Witelonis Perspectivae liber quintus*. Wrocław: Polish Academy of Sciences, 267 pp.

Smith, A M, trans. 1996. *Ptolemy's Theory of Visual Perception: An English Translation of the* Optics *with Introduction and Commentary*. Transactions of the American Philosophical Society 86(2). Philadelphia: American Philosophical Society Press, 300 pp.

Smith, A M, ed. and trans. 2001. *Alhacen's Theory of Visual Perception: A Critical Edition, with English Translation and Commentary, of the First Three Books of Alhacen's* De Aspectibus, *the Medieval Latin Version of Ibn al-Haytham's* Kitāb al-Manāẓir. Transactions of the American Philosophical Society 91 (4 and 5). Philadelphia: American Philosophical Society Press. Vol. I: pp. 3–337; Vol. II: pp. 339–819.

Smith, A M, ed. and trans. 2006. *Alhacen on the Principles of Reflection: A Critical Edition, with English Translation, Introduction, and Commentary, of Books 4 and 5 of Alhacen's* De aspectibus, *the Medieval Latin Version of Ibn al-Haytham's* Kitāb al-Manāẓir. Transactions of the American Philosophical Society 94 (2 and 3). Philadelphia: American Philosophical Society Press. Vol. I: pp. 3–288; Vol II: pp. 289–697.

Smith, A M, ed. and trans. 2008. *Alhacen on Image-Formation and Distortion in Mirrors: A Critical Edition, with English Translation and Commentary, of Book 6 of Alhacen's* De aspectibus, *the Medieval Latin Version of Ibn al-Haytham's* Kitāb al-Manāẓir. Transactions of the American Philosophical Society 98 (1 and 2). Philadelphia: American Philosophical Society Press. Vol. I: pp. 3–153; Vol. II: pp. 155–393.

Smith, A M, ed. and trans. 2010. *Alhacen on Refraction: A Critical Edition, with English Translation and Commentary, of Book 7 of Alhacen's* De aspectibus, *the Medieval Latin Version of Ibn al-Haytham's* Kitāb al-Manāẓir. Transactions of the American Philosophical Society 100(1 and 2). Philadelphia: American Philosophical Society Press. Vol. I: pp. 3–211; Vol. II: pp. 213–550.

Smith, A M 2015. *From Sight to Light: The Passage from Ancient to Modern Optics.* Chicago and London: The University of Chicago Press, 456 pp.

Unguru, S, ed. and trans. 1977. *Witelonis Perspectivae liber primus.* Studia Copernicana 15. Wrocław: Polish Academy of Sciences Press, 330 pp.

Unguru, S, ed. and trans. 1991. *Witelonis Perspectivae liber secundus et liber tertius.* Studia Copernicana 28. Wrocław: Polish Academy of Sciences, 378 pp.

5 Ibn al-Haytham
The Founder of Scientific Pluralism

Hassan Tahiri

CONTENTS

Ibn al-Haytham: Demystifier of the Copernican Revolution 53
Duhem's Interpretation of the Arabic Astronomical Tradition 55
Logical-Epistemic Basis of Astronomical Pluralism 58
References 62

IBN AL-HAYTHAM: DEMYSTIFIER OF THE COPERNICAN REVOLUTION

One of the main aims of this paper is to improve our understanding of the history of science and historical studies. Regarding the development of some scientific disciplines, what we sometimes know or we believe we know is not what actually happened. The modern organisation of knowledge and the structure of the teaching and research little help narrow this gap between science as it is actually made and the dominant perception of its development. The history of astronomy, which is taken as an example, is one of the most studied scientific disciplines. Yet there is no comprehensive work on the history of its development because of the significant gaps which are still not filled. It is a project which has not yet sufficiently drawn the attention of the scientific community, because such work requires an intercultural and collective effort. This is why I refer to the pioneering work of the nineteenth century French historian to illustrate my point.

Light was chosen to be celebrated in the form of an International Year (2015) because of the recognition of the importance in our modern life of light and light-based technologies by the international community; in doing so, it has discovered that the founder of this scientific and technological revolution is Ibn al-Haytham (d. 1040). The celebration of his life and work is a very good initiative that goes in the right direction. For what we are celebrating is not just an Arab-Islamic scientist who lived in the eleventh century but, what is more important, one of those innovative thinkers whose work has reinvented knowledge and greatly contributed to its universalisation and globalisation in his time.* This is one of the significances of his *Kitāb al-Manāẓir* which marks a turning point in the history of science. Thanks to him, optics has finally become a scientific discipline accessible to the rest of the world due to the soundness of its foundation, which has set off its development both in the East and the West, and we hope that this opportunity will encourage researchers to focus on his other works in order to better appreciate the role he played in the transformation of science. Astronomy is the second fundamental discipline that he succeeded in transforming; however, he is less known as an astronomer, though the number of his astronomical treatises is more than twice what he wrote on optics. It is not by chance that, as we will see, his contributions in astronomy are quite as fundamental and founding of our modernity as those of optics. Ibn al-Haytham's interest in astronomy has been neglected, giving the impression that his astronomical work is hardly innovative, although only 3 or 4 at most among at least 25 writings are studied. If it is so, why has our author devoted so much time and energy to this discipline if he did not believe that he could bring something new? History has proved him right; historians, who are only interested in scientific theories as a finished product, underestimate one of the hardest acts of scientific practice: change of scientific tradition. The system of Copernicus has always been considered by modern historians as the absolute beginning of modern astronomy and science. We now know that his astronomy is part of a research programme that was designed to build a non-Ptolemaic system since the eleventh century. This historical fact demystifies the Copernican Revolution since the question of its emergence can now be answered by purely scientific considerations: how did non-Ptolemaic astronomies appear? Or, more precisely, why did research on the construction of non-Ptolemaic systems start in the eleventh century and not before? The answer is provided by the last work of Ibn al-Haytham on the *Almagest* that has changed the literature on the subject: *al-Shukūk* or *Doubts about Ptolemy*, a major change which was not noticed by modern historians except Duhem, but why Duhem? The French physicist especially distinguished himself as a historian by trying to be as comprehensive as possible in his account of the development of science; he produced an unmatched work which enabled him to discover the importance of a tradition that most historians would have found hard to recognise. His approach is at odds with what was followed by most of his contemporaries and successors who have erected an impenetrable wall between science that was developed before and after the so-called Renaissance. Although he is an exception to the prevailing conception of historical studies, his work has exerted a great influence both on history and philosophy. Duhem the historian has indeed exerted a great influence on the philosophy of science. The nature of the relationship between what I call Duhem the historian and Duhem the philosopher is difficult to specify. Several texts suggest that his philo-

---

* It is during the Arabic period that the universalisation and globalisation of knowledge was first achieved. By universalisation of knowledge we mean the emergence of the transnational, transcultural conception of science. The significance of the Arabic-Islamic tradition lies in reinventing knowledge that set off the unstoppable process of de-Hellinisation of the Greek complex legacy which led to the universalisation of science. Ibn al-Haytham's revolutionary work in optics and astronomy can be seen in new light, the so-called Scientific Revolution is part of Ibn al-Haytham's Scientific Program (Rashed's paper in this volume) that inaugurates the irreversible de-Hellinisation movement of natural philosophy using mathematics; for more on this topic see our introduction in Tahiri 2015.

sophical stance comes from his analysis of historical facts; other writings, however, give the impression that he seeks to justify his philosophical doctrine by showing that it is the actual path taken by science throughout history. The French philosopher is known, among other things, for his instrumentalism that he defended with vigour in his famous work: *La Théorie physique, son objet sa structure (Physical theory, its aim its* structure), and by physical theory he means Newtonian physics considered as a complete and perfect system*; his instrumentalism also provides the underlying structure of his little historical book, an extremely compact work of his monumental *Le System du monde*. In *Sauver les apparences (Saving the appearances)* presented as a corroboration of his philosophical doctrine expounded in *La Théorie physique,*† astronomy is not examined for itself but as a chapter of Newtonian physics. It is the same for Kant who took Newtonian physics and not astronomy as a model par excellence of rationality, His *Critique* depends on a monotonic conception of the development of science according to which once a scientific discipline is established, it is established once and for all. Let us follow Duhem's analysis of the historical facts that were available to him to understand how he was able to discover a century later the impact of *al-Shukūk* without knowing it. Overshadowed by Duhem the philosopher, we will see how Duhem the historian stops short of drawing the following conclusion: research on non-Ptolemaic astronomies have started since at least the twelfth century.

DUHEM'S INTERPRETATION OF THE ARABIC ASTRONOMICAL TRADITION

What is intriguing about Duhem is not so much that he is both a historian and a philosopher as to the fact that he seems to have a double interpretation of the history of science that differs according to whether he is a historian or a philosopher, a double interpretation which is determined by his double attitude towards the Arabic astronomical tradition. The philosophical interpretation, which dominated his writings and therefore became more known, explains the emergence of the Copernican system as the result of an evolution of a conflict that took its origin in Greek philosophy: 'alongside the method of astronomy, so clearly defined by Plato, Aristotle admits the existence and legitimacy of [another] method; he calls it the method of the physicist'‡

---

* Duhem seems even to defend a preformationist conception of science. Like a biological process, it looks as if science spontaneously evolves without interruption, change or a subject.

   Those who have of the nature and history of physical theories a deeper view know that to find the seed of this doctrine of universal gravitation, it must be sought among the systems of Hellenic science; they know the slow metamorphosis of this seed during its millennium evolution; they list the contributions of each century to the work that will receive from Newton its viable form; they do not forget the hesitations and groping through which even Newton went through before producing a finished system (Duhem 1993, p 338)

† As indicated by the full title of the work: 'ΣΩΖΕΙΝ ΤΑ ΦΑΙΝΟΜΕΝΑ. Essai Sur la notion de théorie physique de Platon à Galilée'.

‡ By 'physicien' Duhem undoubtedly means natural philosopher since for the Stagirite mathematics has little to do with natural philosophy. The importance of this quotation is that it provides the framework of the whole discussion of the topic. We will show how Ibn al-Haytham's work, which systematically links mathematics with the study of natural phenomena, sheds new light on the intrinsic dynamic of science and consequently provides a more accurate description of the history of astronomy and of science in general.

(Duhem 1908, p. 4). When he comes to the Arabic period, our philosopher argues that this conflict has deepened by emphasising a sharp contrast between the two conceptions of astronomy:

> The Arabs did not share the prodigious geometric ingenuity of the Greeks; they have not further experienced the accuracy and certainty of their logical sense. They only brought slim improvements to the assumptions by which Hellenic Astronomy had managed to solve by means of simple movements the complicated motion of the planets. And on the other hand, when they examined these assumptions, when they attempted to discover their true nature, their views were not equal in penetration to those of Posidonius, Ptolemy, Proclus or Simplicius; slaves of imagination, they wanted to see and touch what Greek thinkers had declared purely fictional and abstract; they wanted to turn eccentrics and epicycles that Ptolemy and his successors considered as artifices of calculation into solid spheres rolling in the heavens; but, even in this work, they have merely copied Ptolemy* (Duhem 1913, volume II pp. 117–118).

This text is a typical example that illustrates the two types of attitude: Duhem the historian has noticed a significant change in astronomical practice during its evolution from the Greek to the Arabic era; Duhem the philosopher classifies and defines them by distinguishing two types of approach to astronomy: the Greek approach, which is presented as instrumentalist or non-realist; and the Arabic approach, which is portrayed as realist; and this is not any realism; it's a naive realism since he speaks of 'seeing and touching objects invented by Greek imagination'. Why does Duhem apparently feel the need to downplay Arabic scientific practice? Let us point out that our philosopher is not quite impartial. In the name of instrumentalism, Duhem was fiercely opposed to the atomist theory and the atomists against whom was directed his *La Théorie physique* and he wanted us to believe that it is the non-realist camp who triumphed at the end of this long battle without mercy by concluding: 'In spite of Kepler and Galileo, we believe today, with Osiander and Bellarmine, that the assumptions of physics are just mathematical devices designed to save the phenomena' (Duhem 1908, p. 140), a conclusion that wants us to believe that realism, despite leading its supporters to the development of classical physics as he admitted, is just a mere historical curiosity.

Along his philosophical interpretation, Duhem has developed a historical description entirely ignored by his posterity; his arguments have convinced neither his contemporaries nor his successors who attempted to overshadow his monumental work. As a historian, Duhem sometimes draws different conclusions from those of the philosopher; let us consider his description of the state of astronomy in the early Arabic period:

> To find an author who discussed the nature of the mechanisms designed by Ptolemy, we need to travel a long period of time and arrive at the end of the ninth century. At that

---

* Unfortunately for Duhem, history has proved him wrong especially regarding the last claim. When he was writing these lines, he obviously did not know of Ibn al-Haytham's *al-Shukūk* nor of the works of the Marāgha astronomers. Because of his limited knowledge of the Arabic-Islamic tradition, he was not able to recognise that what he calls Arabic realism is the result of a new conception of science that binds theory to action.

time, the scientist and prolific astronomer Sabian Thabit ibn Kourrah wrote a treatise in which he strove to assign to the heavens a physical constitution that would accord with the Ptolemaic system (Duhem 1908, pp. 28–29).

This passage is very significant for a number of reasons:

The long stagnation of astronomy which did not move one step forward since the second century.*

The criticism of the Ptolemaic system comes from within, that is, from astronomers and not traditional philosophers, which is a major new development.

Criticism began as the *Almagest* was translated by correcting some essential astronomical parameters for mathematical astronomy.

Ptolemy, who presented his system as a fait accompli, has effectively put an end to the Greek philosophical debate on the subject of astronomy. This means that the fate of his system is in the hands of his successors for a very simple reason: to criticise Ptolemy is to act against him; it is to build something new. This idea, which today appears so obvious to us, was not always the case and it took a long time to see the light of day. Duhem finds it three centuries later in al-Bitrūjī, a twelfth century Andalusian astronomer; from the work of Thābit ibn Qurra to that of al-Bitrūjī, he discovered that astronomy underwent a radical transformation that he described in the following terms:

There is a proposition which can be formulate without reservation and the rest of this writing will justify: this work [of al-Bitrūjī] which is only an attempt and that does not end will have the greatest influence on the evolution of Western astronomy. This influence, we will recognise it everywhere and forever, going alongside the one that exerted by the doctrine of Ptolemy, opposing and preventing it from winning the unanimous acquiescence consent of astronomers. The perpetual conflict of these two influences will maintain the doubt and hesitation with respect to each of them; it will not allow intelligence to be enslaved by the undisputed empire of one or the other of them; it will provide to inquiring minds the freedom of research without which the discovery of a new astronomical system was remained impossible (Duhem 1913, volume II, p. 171).

This proposition is quite special and what makes it so special is when he says 'this writing will justify', to whom will he justify? Our historian seems to have discovered something important to which he wanted to attract the reader's attention; it is thus a message addressed more specifically to his colleagues; what is this message? Let us point out some of his keywords: perpetual conflict, freedom of

---

* The renewal of scientific studies in the Arabic period and their unprecedented development since the ninth century was taken for granted since no scientific explanation for their stagnation is provided. Failure to address this important issue gives the misleading impression that the major gap is not due to epistemic considerations. As a result, many modern historians including Duhem were unable to recognise that the emergence of what they call modern science requires a modern epistemic attitude to purify and free knowledge from Greek cultural and metaphysical conceptions, this is the process that we called the de-Hellinisation of the Greek complex legacy set off by the advent by the Arabic-Islamic civilisation.

research, new astronomical system, impossible discovery. These terms indicate the dynamic nature of his proposition. Duhem completes it by reminding us of the scientific context in which this new kind of research appeared: 'according to Averroes and Al Bitrogi, Abu Bekr ibn Tofail went farther; he tried to build an astronomy where epicycles and eccentrics were also banished' (Duhem 1908, p. 32). These last two passages show how close the historian is to formulate a major thesis: by 'farther' he means 'farther than Ibn Ṭufayl's predecessors' which suggests that this is a new fact; the following sentence tells us more about this novelty, 'an astronomy where epicycles and eccentrics were banned'. Duhem has made no attempt to characterise this astronomy so as to link it to his proposition which describes the new start it took until Copernicus. The description is there, the idea is present but the key concept that he needs to transform his long proposition to a concise thesis is absent. What has created the perpetual conflict between the two systems of the world during this period that Duhem was able to perfectly describe is the attempt to build a non-Ptolemaic astronomy: a powerful logical-epistemic concept which extends to include the work of Copernicus, it is also an epistemic criterion of demarcation between two periods. Duhem the historian could therefore have shifted his focus had he concluded: research on non-Ptolemaic astronomies has begun since the twelfth century. Why was he not able to forge this concept which captures a process of transformation quite specific to scientific theories? How are we to explain this tension between Duhem the philosopher and Duhem the historian? In part by major gaps that Duhem the historian could not fill, the scientific facts that the author of *Le système du monde* did not know are of such magnitude that he could in no case guess them:

1. The new astronomy of al-Bitrūjī belongs to the reform movement triggered by Ibn al-Haytham's *al-Shukūk*.
2. In the east, Ibn al-Shāṭir, one of the Marāgha astronomers, developed the first non-Ptolemaic system in the fourteenth century.
3. Two centuries later, Copernicus provided a second model; his system is derived from and thus equivalent to that of Ibn al-Shāṭir.

In the following section, I will discuss some of these historical facts which could have enabled Duhem, had he known them, to correct his philosophical views as he remarkably admitted: 'Whenever the mind of a physicist is about to commit some excesses, the study of history straightens it by an appropriate correction' (Duhem 1993, p. 411). We will see in particular that the transformation of astronomy is due to the emergence of a new type of logical-epistemic reasoning unknown to the Greeks.

LOGICAL-EPISTEMIC BASIS OF ASTRONOMICAL PLURALISM

Al-Shukūk is a landmark in the history of science since it represents the last work of a scientist, Ibn al-Haytham, both on the *Almagest* and all of Ptolemy's scientific works. It is also a particular work because of its object and structure. Ibn al-Haytham explains his doubts one by one. It is he, as an opponent, who expounds the refutation by choosing the most significant passages where he considers that the proponent is neither convincing nor credible: each of Ptolemy's arguments is followed by a response from his opponent. As a matter of illustration, I present his doubt on Ptolemy's moon model. Ptolemy manages to account for the movement of the five planets, including the moon, by introducing increasingly complicated assumptions* whose common feature is that they contradict his own belief in the principle of circular uniform motion,† and they can therefore be called Ptolemaic assumptions (Figure 5.1). Ibn al-Haytham rejected these hypotheses in two stages: their refutation followed by proof of their absurdity.

Ibn al-Haytham's refutation of the Ptolemaic assumptions is simple and sharp.

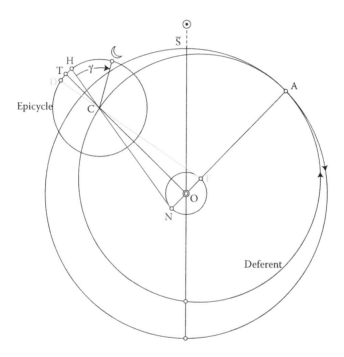

FIGURE 5.1   Ptolemy's model for the motion of the moon. F is the centre of the deferent; C is the centre of the epicycle; FCD is the axis of the deferent. The centre of the epicycle C, which moves the moon, revolves around N known as the prosneusis; N in turn constantly moves so as to remain opposite to the moving centre F of the deferent that is, NCH (in red) moves without passing through the centre.

---

* To account for the motions of the superior planets, Ptolemy introduced the notion of the equant, according to which a sphere moves uniformly around an axis without passing through its centre. In the case of mercury and the moon, their models have the particular feature that the equant point is not fixed.

† Ptolemy has nowhere admitted that his assumptions contradict his belief in uniform circular motion. It is Ibn al-Haytham who eloquently exhibits this contradiction in Ptolemy's belief by successively quoting the two contradicting statements (pp. 23–24). Ptolemy's failure to recognise his contradiction has serious consequences on the development of astronomy and on the analysis of its history.

By definition: a moving sphere revolves around its axis that passes necessarily through its centre F, that is, its diameter (the blue line). He notes that, according to Ptolemy's complex theory of the movement of the moon, it is NC (the red line) which is in motion and concludes that the diameter NC and thus the extending line NH is an imaginary line. This conclusion can in no way worry Ptolemy for he claims in some passages that his concepts are imaginary entities and, as a result, they should not be confused with terrestrial bodies for they have no reality. The situation dramatically changes when Ibn al-Haytham (1971) adds:

> And the diameter of the epicycle is an imaginary line and an imaginary line does not move by itself with a sensible movement that produces something existing in the world* (p. 15).

It is in this passage where Ibn al-Haytham further clarifies the structure of a physical theory by accurately describing the complex machinery of the *Almagest*:

> He has gathered all the motions that he could verify from his own observations and from those of his predecessors. Then he sought a configuration of real existing bodies that exhibit such motions, but could not realise it. He then resorted to an imaginary configuration based on imaginary circles and lines, although some of these motions could possibly exist in real bodies (p. 41).

Ibn al-Haytham considers the *Almagest* as a mixed theory because motion, which is a physical notion, divides its concepts into two classes: (1) physical entities which are provided with a circular uniform motion, these entities are abstracted from real existing objects and their properties are real properties since they are possessed by real existing objects; (2) the imaginary entities, however, are unable to acquire the physical property of motion.

To demonstrate the unintelligibility of the Ptolemaic hypotheses, Ibn al-Haytham did something that Ptolemy should have done: to examine their soundness. Since a spherical body cannot by itself move around a point other than its centre, he considers the possibility that the imaginary movement can be produced by an external force exerted by another body. In a long argument, he shows that such a body should not only perform two types of movement, but that these two movements should also be carried out in two opposite directions. He concludes: 'now this is an absurd impossibility: I mean that one and the same body should possess two opposite, natural and permanent motions' (p. 19). This means that the movement of the imaginary entities cannot be justified. Ibn al-Haytham arrived at what I call the paradox of the *Almagest*: Ptolemy seems to have succeeded in accounting for the movement of the planets by supposing the movement of fictional objects. After analysis, Ibn al-Haytham takes three major actions. The first is the rejection of Ptolemy's system which amounts to a break with his tradition:

> But if one imagines a line to be moving in a certain fashion according to his own imagination, it does not follow that there would be a line in the heavens similar to the

---

* Unless indicated otherwise, all Ibn al-Haytham's quotations are taken from his *al-Shukūk*.

one he had imagined moving in a similar motion. Nor is it true that if one imagined a circle in the heaven, and then imagined the planet to move along that circle, that the [real] planet would indeed move along that circle Once this is established, then the configuration imagined by Ptolemy for the five planets is false (pp. 41–42).

Then follows this revolutionary declaration:

The truth that leaves no room for doubt is that there are correct configurations for the movements of the planets, which exist, are systematic, and entail none of these impossibilities and contradictions, but they are different from the ones established by Ptolemy (pp. 63–64).

To understand to what extent this statement is revolutionary is that after the rapid spread of *al-Shukūk*, everyone, philosophers and astronomers, embarks in search of this new astronomy. How to explain this resounding success of *al-Shukūk*? By a third epistemic act:

The seeker after truth is… the one who follows proofs by argumentation (الحجّة) and demonstration (البرهان) rather than the assertions of a man whose natural disposition is fraught with all kinds of imperfection and shortcomings (p. 3).

Al-Shukūk is not a rhetorical but an argumentative work as underlined by the author. Contrary to rhetoric, argumentation aims at action; he ends his introduction by announcing his research programme:

When we examine the writings of… Claudius Ptolemy, we find in them obscure passages, outrageous terms, and contradictory notions… We find that first and foremost we have to mention their places, and to make them explicit for one who afterwards applies himself to filling the gaps and correcting these notions by any means susceptible to lead to the truth (p. 4).

By 'one who applies himself', Ibn al-Haytham thinks first of himself. In fact, this passage already discloses his next work *The Configuration of the Movements* (in Rashed 2006) in which he tried to work out his new astronomy. I was always amazed by the fact of how a little work, which is about a tenth of the size of Ptolemy's *Almagest*, can change the course of astronomy forever. This is no longer the case today; in fact, it is quite the contrary; it would be very surprising if *al-Shukūk* had not had such a decisive impact when we consider the sophisticated means he used to change once and for all an old tradition from within: a first in the history of science. By putting an end to the Ptolemaic tradition and announcing a new one, he simply reinvented astronomy by indefinitely propelling scientific research to explore new unforeseen ways to understand the world. From the eleventh to the sixteenth century, we went from the belief in the eternal existence of a single model of the planetary system to the coexistence of three models at the time of Copernicus competing between them not to save the appearances but to offer a better explanation of the world: that of Ptolemy imagined in the second century, that of Ibn al-Shāṭir formulated in the fourteenth century and that of Copernicus which appeared in the sixteenth century.

Unlike our instrumentalist philosopher who keeps repeating in his various works that conformation to observations is the only condition to which a physical theory should comply, *al-Shukūk* shows us that this criterion, while necessary, is by no means sufficient since the epistemic subject always reserves the right to refute the established scientific theory. Its author even argues that a scientist should refute it by devising a better model; the reason is of a logical-epistemic character.

> We imagined some arrangements that are adequate for celestial motions. If we imagined other arrangements that are suitable for these movements, there would be nothing against it. Because it has not been demonstrated that no proper arrangement for these movements is possible other than the first (in Al-Bayhaqī 1946, p. 87).

In *al-Shukūk*, Ibn al-Haytham discovers one of the important means of progress underlying the emergence of scientific theories: a scientific theory is based on assumptions that can in no way force us to accept; if we are not convinced by the soundness of one of them, we should reject it and try to develop a better theory based on more satisfactory assumptions. Alexander Koyré, one of the most influential successors of Duhem the historian, regards the Copernican Revolution as the trigger that sparked the Scientific Revolution, and finds that it consists in man's conceptual transformation of the world from a closed to an infinite universe. We now know that this cosmic revolution arises in fact from an epistemic revolution that precedes it and makes it possible: scientific pluralism founded by the author of *al-Shukūk*.

REFERENCES

Al-Bayhaqī, D: 1946, *Tarikh Ḥukama' al-Islam*, al-Taraqqi Press, Damascus, 203 pp.

Duhem, P: 1914, *Le Système du monde: Histoire des doctrines cosmologiques de Platon à Copernic*, Tome II, Hermann, Paris, 522 pp.

Duhem, P: 1993, *La théorie physique. Son objet-Sa structure*, Vrin, Paris, 524 pp.

Duhem, P: 1908, ΣΩΖΕΙΝ ΤΑ ΦΑΙΝΟΜΕΝΑ: *Essai sur la notion de théorie physique*, Vrin, Paris, 140 pp.

Ibn al-Haytham: 1971, *Al-Shukūk 'ala Ba.lamyus* (Edition of the Arabic text by Sabra and Shehaby), Dar al-Kutub, Cairo, 92 pp.

Rashed, R: 2006, *Les mathématiques infinitésimales du IXe au XIe siècle, Volume V: Ibn al-Haytham. Astronomie, Géométrie Sphérique et Trigonométrie*, al-Furqān Islamic Heritage Foundation, London.

Tahiri, H: 2015, *Mathematics and the Mind. An Introduction into Ibn Sīnā's Theory of Knowledge*, Springer Briefs in Philosophy. Springer, Cham, Heidelberg, New York, Dordrecht, 76 pp.

6 Ibn al-Haytham
Founder of Physiological Optics?

Vasudevan Lakshminarayanan

CONTENTS

Introduction 63
The Oeuvre of Ibn al-Haytham 65
Visual Psychophysics and al-Haytham 67
Anatomy of Eye and Schematic Eye Models 67
Colour 73
Dark and Light Adaptation 78
Directional Sensitivity 79
Eye Movements 80
The Horopter and Binocular Vision 82
Law of Equal Innervation 87
The Moon Illusion 88
Motion Perception 91
Pattern Classification and Object Recognition 93
Size Perception 96
Unconscious Inference and Perception 98
Visual Acuity 101
Conclusion 103
References 104

INTRODUCTION

The field of optics, which is the study of the science of light (or, more generally, electromagnetic phenomena) and vision, can be divided into four main areas: (1) geometric optics, (2) physical or wave optics, (3) quantum optics and (4) physiological optics.

What is physiological optics? The principles governing the eye's perception of light is the area studied by the science of physiological optics (or vision science in more modern parlance). Physiological optics is the investigation of the dioptrics of the eye, and the visual system from the eye to the cerebral cortex of the brain and incorporates biophysics, neuroscience, psychology and cognitive science, as well as computational and theoretical approaches to study visual perception.

According to the United States National Center for Education Statistics, a pro
gramme of study in physiological optics is defined as

> A program that focuses on the scientific study of vision, visual processes, and related
> phenomena and clinical research and treatment modalities. Includes instruction in ocu-
> lar anatomy and physiology, microbiology of the eye, electrophysiology, neurophysiol-
> ogy, corneal physiology, photochemistry, psychophysics, visual biophysics and motor
> systems, sensory mechanisms and photoreception, eye circulation and metabolism, geo-
> metric and physical optics, ocular development across the life span, visual stimuli and
> perception, color vision, eye motility, biometrics and measurement techniques, visual
> pathology, and environmental issues (National Center for Education Statistics, 2016).

The Oxford English Dictionary gives a compact definition of the noun, physiological optics, as 'The scientific study of the perceptual and physiological processes
involved in sight'. The OED also claims that the first use of the term was in the
Proceedings of the Royal Society in the mid-nineteenth century.

Even though many luminaries of science (and optics) such as Kepler, Scheiner,
Newton, Huygens and Young had contributed to various aspects of the science
of vision, the first comprehensive, magisterial text was written by Herman von
Helmholtz (1821–1894). Helmholtz was a polymath, who made fundamental contributions to optics and physics (e.g., the Helmholtz equation; the conservation of
energy, thermodynamics, acoustics) and the psychophysics of sound, physiology,
psychology and philosophy (Koenigsberger, 1965 for a fascinating biography).
Helmholtz, in addition to revolutionising the field of ophthalmology by his invention
of the ophthalmoscope, went on to systematise the study of vision in the three volume
Handbuch der Physiologischen Optik (*Handbook of Physiological Optics* or *Treatise
on Physiological Optics*). The *Handbuch* was translated into English by James P
C Southall on behalf of the Optical Society of America (Helmholtz and Southall,
1910/1962) in 1924–1925. In this landmark book, Helmholtz discussed empirical
data and theories on various aspects of vision, including color vision, motion perception, depth perception, spatial vision, eye movements and accommodation (in fact,
his theory of accommodation was challenged only in the decade of the 1990s by
a capsular theory; Schachar, 1994), as well as on dioptrics of the eye. In the third
volume, he came up with the idea of unconscious inference ('Unbewusster Schluss';
unconscious conclusion is the term used by Southall) for perception and discussed its
importance (Hatfield, 2002 for a good discussion). This is a major idea of cognition
and perception Even though Helmholtz did not coin the term Physiological Optics, it
will forever be associated with him and many consider him the founder of the field
of physiological optics.

What was Ibn al-Haytham's role in this field? Ibn al-Haytham used ideas of human
eye anatomy and of vision from the Greeks. However, he enlarged and expanded the
knowledge base by carrying out various experiments on vision and by accurately
describing the results and conclusions to be drawn from them. Al-Haytham was
probably the first to underscore the fact that vision does not end with the formation
of the image by the optics of the eye. Though Galen and others had speculated that
vision involves the brain (Lindberg, 1976), al-Haytham was the first to explicitly
state the role played by the brain. He devoted Book II of his magnum opus Kitab

al-Manazir to perception. It appears as though he described or anticipated many visual and perceptual phenomena that were (re-)discovered in Europe (and elsewhere) centuries later.

In this chapter, I assert that even though there are other contenders to the title of founder of physiological optics, Ibn al-Haytham should be considered the true founder. I discuss some of the many contributions of Ibn al-Haytham to the study of vision and visual/perceptual phenomena which justify this assertion.

THE OEUVRE OF IBN AL-HAYTHAM

Ibn al-Haytham was a polymath and a prolific writer who is reputed to have written over two hundred works on a wide range of subjects ranging from optics, vision, number theory, mathematics, geometry, meteorology and astronomy to medicine and philosophy. Ninety-six of his scientific works are known. Most of his works are now lost, but more than 50 of them have survived to some extent. Nearly half of his surviving works are on mathematics, 23 of them are on astronomy and 14 of them are on optics, along with a few on other subjects (Rashed, 2002). Not all his surviving works have yet been studied, but some of the ones that deal with optics and vision include: (1) the *Kitab al-Manazir (Book of Optics)*, (2) *On the paraboloidal burning mirror*, (3) *The spherical burning mirror*, (4) *On the burning sphere*, (5) *On light*, (6) *On the rainbow and halo*, (7) *On the nature of shadows*, (8) *On the form of the eclipse*, (9) *On the light of the moon*, (10) *On the light of the stars*, (11) *Doubts concerning Ptolemy*, (12) *An analysis of the optical knowledge of Euclid and Ptolemy*, (13) *On the nature of sight and the manner in which sight occurs*, (14) *On optics according to Ptolemy*, (15) *An analysis of Aristotle's De anima,* (16) *An analysis of Aristotle's Meteorologica* and (17) *On the perfection of the art of medicine.* This last book contains chapters on diseases of the eye. Another book *'On the twilight'* is frequently attributed to al-Haytham. A complete listing of all of his works, manuscripts, editions and translations is given by Sabra (1972; pp. 204–209). The article by Sabra (1972) also gives a good overview of al-Haytham's life (Sa'di, 1956).

What is even more noteworthy is that al-Haytham is often called the first theoretical physicist. In mediaeval Europe, he was known as 'the physicist'. Many historians of science consider Ibn al-Haytham to be the first true proponent of the modern scientific method (Al-Khalili, 2015). He firmly laid down the basic tenet of the scientific method centuries before Roger Bacon – that is, a hypothesis must be proved by experimental observations or mathematical evidence. According to al-Haytham (Sabra, 1989):

> The seeker after truth is not one who studies the writings of the ancients and, following his natural disposition, puts his trust in them, but rather the one who suspects his faith in them and questions what he gathers from them, the one who submits to argument and demonstration and not the sayings of human beings whose nature is fraught with all kinds of imperfection and deficiency. Thus the duty of the man who investigates the writings of scientists, if learning the truth is his goal, is to make himself an enemy of all that he reads, and, applying his mind to the core and margins of its content, attack it from every side. He should also suspect himself as he performs his critical examination of it, so that he may avoid falling into either prejudice or leniency.

It is appropriate to quote the great American physicist Richard Feynman (1918–1988; 1985 [1974]) 'Science is what we have learned about how not to fool ourselves about the way the world is'.

Al-Haytham's optics and vision research methodically and systematically relied on experimentation under controlled conditions. In fact, according to the science historian Matthias Schramm, al-Haytham 'was the first to make a systematic use of the method of varying the experimental conditions in a constant and uniform manner' (Toomer, 1964). Until recently, historians of science often used to credit Galileo Galilee with the birth of modern science, in particular the emphasis on experimentation and use of mathematics. However, it is clear that Ibn al-Haytham is rightly the founder of modern science. The Italian historian Gorini (2003) correctly says, 'According to the majority of the historians al-Haytham was the pioneer of the modern scientific method. With his book, he changed the meaning of the term optics and established experiments as the norm of proof in the field. His investigations are based not on abstract theories, but on experimental evidence and his experiments were systematic and repeatable'.

In addition, he combined the known physics with mathematical geometry to support his hypotheses and laid the groundwork for his work on vision. It was Ibn al-Haytham's faith in empirical data and trust in mathematical proof that was the foundation of the *Kitab al-Manazir*. This book was the guide star for vision science that was initiated in Europe in the centuries following al-Haytham.

The Kitab al-Manazir comprehensively formulates the laws of reflection and presents an investigation of refraction, with experiments involving angles of incidence and deviation. He also explains refraction as being due to the change in the velocity of light in different media. The work also contains what is known as: 'Alhazen's problem' – to determine the point of reflection from a plane or curved surface, given the center of the eye and the observed point – which is stated and solved using conic sections (hyperbolas). Book I largely deals with optics, the structure of the eye, image formation in the eye and with the visual pathways. However, Books II and III deal with the experiments carried out by him on a broad range of visual perceptual phenomena such as the theory of unconscious inference; the law of equal innervation of the eye; the principles of binocular direction; constancy of size, shape and color; induced visual motion; the vertical horopter; the fusional range of binocular disparity, color assimilation and many others. Howard's article (1996) was, to the best of my knowledge, the first to discuss some of these contributions. In this chapter, I concentrate on some of al-Haytham's contributions to Physiological Optics. Unless otherwise indicated, the quotations from the *Kitab al Manazir* are from the English translation by Sabra (Volume 1, 1989).

Visual/perceptual studies are mostly conducted using the methods of psychophysics. I start this chapter with a discussion of al-Haytham and psychophysics. Then I go on to discuss the studies on the anatomy of the eye (Section 'Anatomy of Eye and Schematic Eye Models') and the description of the first schematic eye model. This is followed by (in alphabetical order) various visual and perceptual phenomena that were investigated by al-Haytham.

I do not discuss in detail probably his most significant major accomplishment, namely his proof of the intromission theory of vision. This is covered in great detail elsewhere (e.g., Lindberg, 1967, 1976; Park, 1997). In this chapter, I deal with other aspects of the work of al-Haytham on vision.

VISUAL PSYCHOPHYSICS AND AL-HAYTHAM

In order to study visual perceptual phenomena non-invasively, one utilises the techniques of psychophysics. Psychophysics is the quantitative investigation of the relationship between physical stimuli (in physical space) and the perception caused by the stimuli in the brain in a psychic/perceptual space (these spaces are not usually isomorphic). Psychophysics also refers to the class of methods that can be used to study a perceptual system. The word Psychophysics was coined by the German physicist Gustav Theodor Fechner (1801–1887), who, in his landmark book, 'Elements of Psychophysics' (*Elemente der Psychophysik*, 1860), described the fundamental experimental techniques, conducted laboratory psychophysical experiments and established the methodology still used by sensory scientists. For more details on psychophysics, see, for example, Lu and Dosher (2013) or Gescheider (1997).

Can al-Haytham be given credit as being the founder of psychophysics? Given his reliance on the experimental method, and his work on various aspects of visual perception, one might be tempted to do so. In fact, Khaleefa (1999; Khaleefa and Manna, 2000, 2001) poses that specific question and argues that it should be in the affirmative. He specifically writes in his 1999 paper (p. 23), 'It is our conclusion that Ibn al-Haytham deserves the full title of Founder of Psychophysics as well as Founder of Experimental Psychology. *The Book of Optics* by Ibn al-Haytham in the first half of the eleventh century, and not the Elements of Psychophysics by Fechner in the nineteenth century, marks the official "founding" of psychology because it provides not only new concepts and theories, but new methods of measurement in psychology'.

Another writer, Taha (1990) says emphatically, '… Ibn al-Haytham is the founder of the psychology of vision (p. 273) [and] modern psychophysics (p. 262).' (I should point out that these quotations are from a published abstract but I have been unable to obtain the full paper.)

The most serious discussion of al-Haytham as the founder of psychophysics and experimental psychology comes from Khaleefa. However, many of the points raised by Khaleefa are discussed and refuted by Aaen-Stockdale (2008). The major criticism is that Khaleefa conflates experimental psychology with psychophysics. However, more importantly, there is no support in the available translations of *Kitab al-Manazir* (e.g., Sabra, 1966, 1989; Smith, 2001) that al-Haytham used psychophysics (strictly defined) as quantitative perceptual judgements as a function of stimulus variables, although he DID make qualitative judgements of a variety of psychological visual perceptual phenomena.

ANATOMY OF EYE AND SCHEMATIC EYE MODELS

In the seven volume *Kitab al-Manazir*, the first three volumes are devoted to light and vision. In Volume I, the 5th section is completely devoted to the anatomy and structure of the eye (pp. 55–63). In the earlier sections, he systematically investigated the rectilinear propagation of light, independent of the eye. Similarly, he gave a detailed description of ocular anatomy independent of geometric optics. After this description of the anatomy, he then goes on to detail the functional significance of the eye as an optical system. In this section, therefore, he sets forth the anatomy of the eye, the relationships between various parts of the eye and finally how the whole organ works together as an image forming system. As noted by Russell (1996), al-Haytham's eye is characterised by physical/geometric considerations such as the shape, position and condition of the parts of the eye, with emphasis on interrelationships between these structures and their invariant properties.

The basic description of the eye of al-Haytham is remarkably similar to our modern-day view (Figure 6.1).

Ibn al-Haytham's description of the anatomy of the eye and its various structures can be found elsewhere (Polyak, 1941; Unal and Elcioglu, 2009). However, he was not original in this descriptive anatomy. In fact, he gives credit to others ('all that we have mentioned of the eye's coats and of its composition has been shown and expounded by anatomists in the books on anatomy', p. 62). His description of the system is essentially Galenic (in fact, he says often in his writing 'it is said') and is based on existing work rather than first-hand dissection and observation. Al-Haytham's major contribution to the field is his consideration of the detailed anatomy and then relating it to its function as an optical image forming system. In this section, I will deal with the functional anatomy and the schematic eye model he created.

Al-Haytham first describes the concave inner surface of the cornea which, when continued, would include the lens and form the posterior surface of the lens. The crystalline lens was described by Galen and others as anteriorly flattened. However, al-Haytham gives a description based on the radii of curvature of the anterior and posterior surfaces. He states that the anterior surface is part of a sphere which is larger than that of the posterior surface (Figure 6.2).

The globe determining the anterior surface would enclose the lens and vitreous; the shorter posterior surface would be a curve that would be continuous with the anterior surface of the cornea and would consist of part of the lens and the cornea. These two spheres would intersect at the junction of the ciliary body and the lens.

Using these two spheres, he places the lens unequivocally near the cornea. The center of the eye/center of the large uveal sphere therefore falls behind the lens. With this geometrical arrangement in place, he then defines an optic axis for the eye as a line connecting the centers of these two spheres, a straight line that is perpendicular to the chord of intersection of the two spheres bisecting it at right angles. He further defines this axis as one that passes through the center of the globe of the eye, which also goes through both the gni of the pupil and the center of the funnel of the optic nerve. In other words, all refracting media (cornea, aqueous, lens and vitreous) are centerd on this axis (Figure 6.2). Also when describing the eye and the axis, he often uses surfaces and spheres interchangeably, thus considering the eye longitudinally as well as coronally. This is definitely a pioneering achievement. Al-Haytham formulates this axis since it is crucial to his quantitative description of the dioptrics and image formation by the eye later on. Also, he uses the movement of this axis to

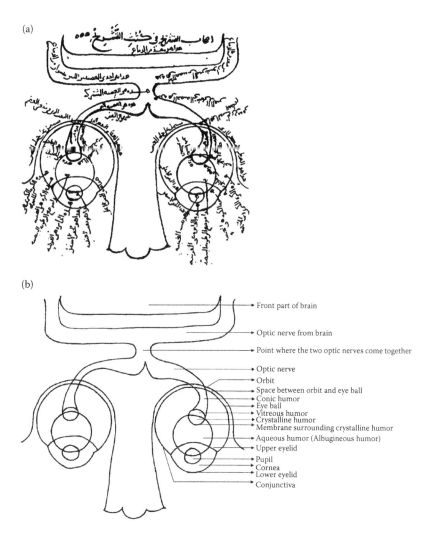

FIGURE 6.1 (a) This is a figure of the human visual system and its relationship to the brain. This figure is from the original Kitab al-Manazir (Faith Library, Fatih Kitapsaray, MS. NO. 312, Istanbul, Turkey). (Modified from S L Polyak, 1941, University of Chicago Press, Chicago, IL.) (b) Parts of the eye as given by al-Haytham and some modern equivalents.

discuss convergent and conjugate eye movements. Of course, a major defect in this model is that the center of the optic nerve lies in the same axis as the center of the pupil and the center of the optic nerve. We know that the optic nerve is ~12–15° nasal to the fovea.

Al-Haytham therefore described the first schematic eye model. However, it should be emphasised that he did not provide values for various parameters such as axial length, curvature, refractive indices, etc. The first 'modern' eye models were developed by Allvar Gullstrand (1909), a Swedish ophthalmologist, for which he received

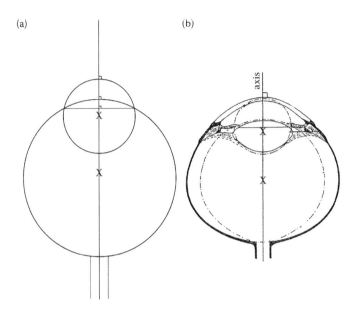

FIGURE 6.2   (a) Schematic eye as developed by Ibn al-Haytham – the central line marks the visual axis; the Xs mark the centers of the two circles. The axis is perpendicular to the center of the cornea. (b) Same as Figure 3a, but superimposed on the eye cross section.

the Nobel prize in Physiology and medicine in 1911. It is important to note, however, that Gullstrand also used spherical geometry to describe various surfaces of the eye and made the assumption that the eye was a centerd optical system with the centers of the different surfaces falling on an optic axis. Many sophisticated schematic eye models have been developed since the time of Gullstrand and are used to investigate the dioptrics of the eye and of image formation and play a very important role in physiological optics (for a review, Smith, 1995).

At this point, we take a slight detour and discuss his contribution to the intromission theory. As noted earlier, detailed analyses are given elsewhere. In ancient times, the problem of how we see had two mutually exclusive theories – the extromission theory and the intromission theory with variants thereof – which coexisted for centuries. In fact, Kuhn (1970) used this as an example of what he called a 'pre-paradigmatic' stage of a scientific theory. According to extromissionism, rays of light are emitted from the eye, light up the object and are sensed by the eye; the intromission theory postulates that rays (or corpuscles of light)from the object entered the eye and produced sensation. Al-Haytham resolved this centuries old issue irrevocably in favour of the intromission theory. In al-Haytham's analysis, what is crucial are the following: (1) A 'punctiform' analysis (Lindberg, p. 30, 1976) – that is, there is a point-to-point correspondence between every object point and each point on the image formed in the eye/retina and (2) The central ray of the visual cone is the most important in conveying perception (section on Directional Sensitivity). The idea of the 'visual cone' is not due to al-Haytham. The credit goes to Ptolemy and the visual cone is defined as the bundle of visual rays centerd in the eye (Lejeune, 1956).

Al-Haytham demolishes the extromission idea by noting that bright lights (and colors) cause eye pain. Therefore, the eye is receiving something from outside and not emitting anything. He uses the idea of after-images and concludes:

> since sight does not perceive any visible object unless it shines, therefore, the form which sight perceives of an object must be according to the light in the object and according to the lights shining on the eye at the time of perception and on the intervening air between the eye and the visible object (p. 55).

and

> when the eye faces a visible object that shines with any light whatever, light comes from the light in the object to the surface of the eye. ...it is a property of light that it affects the sight, and that it is in the nature of sight to be affected by light. It is therefore most appropriate that the eye's sensation of the light that is in the visible object should occur only though the light passing from the object to the eye (p. 63).

He describes in Book I a series of experiments and shows, using logical arguments, that if only light entering the eye gives rise to sensation, then nothing is lost by discarding the idea of extromission, namely light emitted by the eye – these are superfluous (pp. 78–82).

He states categorically

> Therefore nothing issues from the eye that senses the visible object (p. 80).

This set of logical arguments is an outstanding example of the use of Occam's razor!

With regard to the punctiform analysis, al-Haytham correctly stresses that the point-to-point correspondence between the outside world and the representation in the eye has to be maintained and conveyed to the brain. This forms one of the bases of modern visual physiology and neurology. As has been pointed out by Tootell et al. (1982), in the striate cortex, the anatomical substrate for retinotopic organisation is well ordered, and there is a systematic relationship between ocular dominance strips and cortical magnification. Schwartz (1977) has further shown that this retinotopic mapping obeys a log-conformal map; however, this point-to-point analysis is maintained.

Al-Haytham, influenced by Galen and other Arabic scholars such as Ibn Sina and Hunain Ibn Ishak, maintained that the seat of vision is the crystalline humour (or glacialis). He brings in Ptolemy's visual cone in his arguments.

> ...when the eye faces a visible object there is formed between the object and the center of the eye a cone with that center as vertex and the surface of the object as base. There is thus between every point on the object's surface and the center of the eye an imaginary straight line perpendicular to the surfaces of the eye's coats; the cone comprising

these lines will be cut by the surface of the crystalline's surface....the cone will com prise of all those straight lines between the eye and the object though which the form of the object is perceived by the eye. It has also been shown that sensation occurs only though the crystalline. Therefore the eye's sensation of the light and color that are in the surface of the visible object is only through that part of the crystalline's surface which is determined by the cone (p. 83).

He underscores the importance of the crystalline humour and possible causes of vision loss:

We say first that vision must occur through the crystalline,...for vision cannot occur though any of the coats that are anterior to the crystalline (and that are only instruments of the crystalline) for the reason that if some damage befalls the crystalline humor while other coats remain intact vision will be destroyed, but if the other coats are damaged while retaining their transparency (or some of it) and the crystalline remains intact, vision will not be damaged....all this is attested by the art of medicine (p. 66).

Al-Haytham's explanation of how the visual process occurs occupies all of Book I, Chapter 6 'On the Manner of Vision' (pp. 63–100). From a reading of this chapter, it is clear that al-Haytham placed the formation of the visual image on the posterior surface of the crystalline lens. It is this image information that is passed on to the brain through the hollow optic nerve to the anterior portion of the brain:

Now this sensation which takes place at the crystalline extends into the hollow nerve and reaches the front of the brain, where the ultimate sensation takes place. The last sentient, that is, the soul's sensitive faculty resides in the front of the brain. It is this faculty that perceives the sensible objects – the eye being only one of the instruments of this faculty. The most the eye does is to receive the forms of visible objects that occur in it and convey them to the last sentient and it is the latter that perceives those forms and though them perceives the visible properties....The form that occurs in the surface of the crystalline extends into the crystalline's body, then through the subtle body in the hollow of the nerve until it reaches the common nerve. When it occurs there, vision will be effected and from the form occurring in the common nerve the last sentient will perceive the form of the object (pp. 84–85).

However, Polyak (1941, p. 122) argues that al-Haytham's description of the 'gla- cial' sphere occasionally includes both the crystalline lens and the vitreous humour (as we term it today). If the glacialis is interpreted this way, then the visual image is necessarily formed on the posterior surface of the vitreous or practically on the retina.

Finally, we come to the next major inaccuracy of al-Haytham's idea of vision – namely the formation of an inverted image by the optical system of the eye. Al-Haytham, who pioneered the analysis of the camera obscura (Latin for dark room), gave a full account of the principle including experiments with five lanterns outside a room with a small hole. He described the camera obscura in his writings; manuscripts of his observations are to be found in the India Office Library in London (1571). Even though al-Haytham knew an inverted image would occur, it was unacceptable. The inversion of the image would have to be reconciled with the fact that we have a veridical view of the world around us. Therefore, he devised a refractive mechanism at the interface between the crystalline lens and the vitreous and says:

> but if vision is accomplished only when the form which occurs in the crystalline's surface reaches the common nerve with the positions of its parts unchanged; and if this form cannot reach the cavity of the nerve with the positions of the parts unchanged except by being refracted, then vision is not accomplished until after the form which occurs in the crystalline's surface has undergone refraction and extended the lines that intersect the radial lines, this refraction having taken place before the form reaches the center, because if it were refracted after passing through the center it would be reversed (p. 116).

According to al-Haytham, this refraction would result in a right side up image of the visual world to the optic nerve. This problem was not solved until Johannes Kepler (1571–1630) who showed that an inverted image is formed on the retina (Lindberg, 1976; pp. 193–202). It should be underscored that Kepler had to consider non-mathematical aspects to explain how we see an upright non-reversed image.

In spite of these errors, al-Haytham correctly deduced that once the image falls on the light sensitive part of the eye, optical considerations were no longer required. He did emphasise that the point-to-point representation had to remain intact and conveyed to the 'ultimum sensus' in the front part of the brain.

COLOUR

The Islamic world made important discoveries in the field of color science (Table II of Kirchner, 2015). In particular, Ibn al-Haytham made considerable contributions to the role of color in vision. In fact, 'Ibn al-Haytham promoted the role of light for color vision from being a mere catalyst to the very object of sight' (Kirchner, 2015). To Ibn al-Haytham, color was a percept. According to him, light and color are ontologically distinct, but color requires illumination to be seen. He recognized that color was pure sensation and not physical. If color appearance was due to mental/cognitive processes, then this was achieved by the faculty of judgement involving what he called inference (Sabra, 1978).

He says in Book II, Chapter 4: On the effect of light upon sight,

> for the sense of sight perceives the forms of visible objects from the forms that come to it from the colors and lights of these objects. And, it's perception of lights qua lights and of colors qua colors is by pure sensation (p. 130).

A caveat should be inserted here: Over twenty different color names/terms are found in the *Kitab*. Sabra (vol. 2, 1989, pp. 40–44) notes that many of the color terms used by al-Haytham are difficult to interpret for a number of reasons, including translation, and references to unfamiliar objects. Sabra notes that 'questions of the denotation of color terms are more crucial than about the ontological status of color' (p. 40).

Al-Haytham discusses in detail the fact that color is real and inherent in objects and not the result of some sort of effect created by light in the eye (pp. 48–49). In his discussion of color, al-Haytham makes a very important connection between an opacity and color. He states that this is a necessary connection, namely every opaque object has color or in the case of stars and fire, 'something like color':

> but every opaque body has a color or something like color such as the light of the stars and the forms of (self-)luminous bodies. Similarly, no transparent body with any opacity in it can be devoid of color (p. 9).

Having dealt with the nature of color of objects, he then deals with the question of how the visual system deals with color. He says:

> ...perception of color as color takes place before perceiving the quiddity of the color; I mean that sight perceives color and senses that it is color, and the beholder who looks at it knows that it is color before realizing what color it is (p. 143).

To back this up, he further states:

> Thus, perception of color qua color precedes perception of the quiddity of the color, the latter being achieved by recognition. A proof that sight perceives color qua color before perceiving what color it is , is furnished by visible objects of strong colors such as dark-blue, wine and "missani-green" and the like when they exist in a somewhat obscure place. For when sight perceives one of these colors in a dim place, it perceives it only as a dark color, realizing that it is color without discerning what color it is. If the place is not very dim, sight will perceive what the color is after contemplating it further. And it will discern what the color is if the light in that place becomes stronger.

In the last sentence above, Ibn al-Haytham recognizes the fact that in order to have color vision, we need to go to photopic levels of illumination. He further elaborates on the role of illumination:

> the form of the color in the illuminated body is always mingled with the form of light. Thus when the form of light arrives from the object at the eye it affects the latter by virtue of its strength...the eye perceives the color of the object according to the light in the object...and colors of many visible objects change according to variation in the lights shining upon them. And since the form of the color cannot produce an effect on the eye unless it is mixed with light (p. 107).

In his discussion of color, al-Haytham also describes the loss of color in the peripheral visual field:

> if a test is made of a variously colored object ...the object being placed outside the radial axis, and very far from it, the experimenter will find the object to be such as we described it, namely that it is of a single uncertain color. Sight will therefore err in what it perceives of the color of this object (p. 264).

Having discussed the recognition of color and its perception, he then goes on to color mixing. His experimental tool for this is the use of a color wheel/top. The color wheel is a simple and effective way to produce color mixing. This, of course, was not new, however. Earlier, Ptolemy devised a wheel on which different colors could be presented either as dots or in sectors and when the disk is rotated rapidly, the dots would appear as colored circles or the sectors would combine to form a single color (Smith, 1986, pp. 109–110). Ibn al-Haytham uses his version of the color wheel to prove his statement that:

> Perception of the quiddity of color must take place in time. For, perception of the quiddity of color can only be achieved by discernment and comparison. But, discernment must take time and therefore, perception of the quiddity of color must take time (p. 144).

The color top devised by al-Haytham is:

> painted in different colors forming lines that extend from the middle of its visible surface close to its neck to the limit of its circumference then forcefully made to rotate it will turn around at great speed. Looking at it the observer will now see one color that differs from all the colors in it as if this color were composed of all the colors of these lines; he will neither perceive the lines nor their different colors […] But all points at equal distances from the center will move with the top's rotation on the circumference of a single circle. In consequence of this, the color of every one of the points at equal distances from the center will appear on the circumference of one and the same circle in the smallest amount of time, which is the same as the duration of one revolution; therefore, the colors of all those points will appear in the circumference of that circle as mixed and undiscerned by sight; and thus sight will perceive the color of the top's surface as one color that is mixed of all the colors in its surface. […] It is seen from this state of affairs that perception of the quiddity of color must require time (p. 145).

Since the visible colors on the top are not recognisable, the visual system makes an error and hence an illusion. Ptolemy in his experiment with the color wheel essentially concludes that fast color changes cannot be distinguished. However, Ibn al-Haytham states that it takes 'a smallest amount of time' (p. 145). However, contrary to his descriptions of other experiments, this description of color mixing is brief and lacks accuracy. His main conclusion is that when 'a stronger color' is mixed with a 'weaker' color, the resulting mixture is dominated by the stronger color. Unfortunately, we do know if he is referring to color saturation or the size of the areas of the disk with those colors. He does not make any conclusive statement about mixing specific colors. Also, just like in Ptolemy's account, there are no conclusions about mixing particular colors.

In his discussion of color, al-Haytham basically gives a statement of the idea of unconscious inference (Section 'Unconscious Inference and Perception') and goes on to say that the inferential step between sensing color and differentiating it is shorter than the time taken between sensing and any other visible characteristic (except light) and that 'time is so short as not to be clearly apparent to the beholder' (p. 146). This, of course, implies that the color and form are sensed somewhere else.

Al-Haytham states further that in processing the information, the information has to go to the cavity of the common nerve and:

> the sentient organ does not sense the forms that reach it from the visible objects until after it has been affected by these forms; thus it does not sense color as color or light as light until after it has been affected by the form of color or light. Now the affectation received by the sentient organ from the form of color or of light is a certain change; and change must take place in time; …..and it is in the time during which the form extends from the sentient organ's surface to the cavity of the common nerve, and in (the time) following that , that the sensitive faculty, which exists in the whole of the sentient body will perceive color as color…Thus the last sentient's perception of color as such and of light as such takes place at a time following that in which the form arrives from the surface of the sentient organ to the cavity of the common nerve (pp. 147–148).

Next, we turn our attention to the work done by al-Haytham on after-images. After-images due to brief intense light or fixating on an illuminated object were used by al-Haytham to justify the intromission theory of light (see, e.g., Lindberg, 1976, pp. 58–76). Historically, after-images have also been called accidental colors referring to the colored characteristics of the phenomena. After-images, according to Franz (1899) 'have an epitome of the interrelations of physics, physiology and psychology and probably no other single phenomenon is so good an example of the growth of experiment and measurement in psychology'. It is therefore no surprise that al-Haytham did many experiments on after-images in his unification of physics, physiology/anatomy and perception in vision science. He, as one would expect, was very precise. He noted that the shape of the primary stimulus was visible in the after-image and it vanished with time.

> if an observer looks at a pure white body irradiated by daylight so that the light on this body is strong, although it is not sunlight, and he continues to look at this body for some time, then he turns his eyes to a dark place, he will find the form of that light in the same place, and with the same shape.. .When subsequently closes his eyes and stares for a while, he will experience in his sight the form and shape of that light; then all that fades away and the sight returns to its own condition (p. 51).

He then turns his attention to the effects of color on after-images and concludes:

> it is therefore evident from these experiments that illuminated colors also have an effect on sight.

He comes to this conclusion by instructing the reader to observe a verdant green meadow and then a dark place or if they view a white sheet, it will be the color green.

The last topic to be discussed in this section is on color constancy and color contrast. Consider the fact that even though the retinal image is in constant flux, our perception of the world is one that is stable and contiguous. The reason for this is that the visual system seems to be able to extract certain invariances under these transformations. These invariances are called constancies and include size, shape, angle and color. We typically see objects as having the same color (e.g., a red rose as being the same shade of red) under varying conditions of illumination (e.g., in sunlight and under fluorescent lighting which have different spectral distributions). There is, within limits, constancy of perceived color under variations in the intensity and color of ambient light. Color contrast, on the other hand, is the change in the appearance of a color surrounded by another color; that is, our perception of hue depends upon its interaction with adjacent hues; for example, grey looks bluish if surrounded by yellow. A good historical reference on various aspects of color is the book by Wasserman (1978).

Al-Haytham understood that perception of colors depended upon the surroundings. He describes how to see contrasts with paint. The example given is green paint on different backgrounds:

> if designs are made with fresh-green paint on a dark-blue body, the paint will look [red] and of a clear color; but if designs are made with the same paint on a clear-yellow body, the paint will look [green] and of a dark color. And similarly with all paints that are intermediate between the two extremes.

And the importance of contrast:

> Therefore when the colors and lights of neighboring objects are excessively in contrast with one another in respect of strength and weakness, the true nature of the weak among them will not be apparent or perceived by the eye next to those that are strong and contrasting. For the qualities of lights and colors are perceived by the eye only by comparing them with one another. Strong lights hinder the eye from perceiving objects whose lights are weak because the forms of the weak lights mix with those of the strong lights and in these mixtures the forms of the strong become dominant over those of the weak; or because the weak lights are close to the strong, and (because) the eye perceives contiguous and homogeneous forms by comparing them with one another and (because) the sense(-faculty) is unable to perceive what is very weak in comparison with a strongly sensible object (Sabra, vol. I, pp. 99–100).

Again, as noted previously, many of the color terms that al-Haytham uses are difficult to interpret. Strong and weak lights probably refer to the hue. To continue this discussion, let us now see what he has to say about simultaneous contrast (Wade, 1996):

> Thus if a pure white body is dotted with a dark-colored paint by allowing small drops of the paint to fall on it, or if minute designs are made on it with this paint, the paint will look black or very dark; its distinctive quality will cease to be apparent and the eye will not be able to perceive its true color. If marks are made with the same paint on a pitch-black body, the paint will look white or pale-colored; its darkness will not be apparent and the eye will fail to perceive its true color. If however, this paint is placed in midst of bodies that are not extremely white or extremely black, its color will appear as it is and the eye will perceive its true color in so far as it can be perceived by sight ... For the qualities of lights and colors are perceived by the eye only by comparing them with one another (Sabra, vol. 1, p. 99).

Al-Haytham described color constancy by observing that light reflected by an object is modified by the color of the object. He explained that the quality of the light and color of the object are 'mingled' and the visual system separates out the light and color. In Book II, Chapter 3, he writes:

> Again the light does not travel from the colored object to the eye unaccompanied by the color, nor does the form of the color pass from the colored object to the eye unaccompanied by the light. Neither the form of the light nor that of the color existing in the colored object can pass except as mingled together and the last sentient can only perceive them as mingled together. Nevertheless, the sentient perceives that the visible object is luminous and that the light seen in the object is other than the color and that these are two properties (p. 141).

In his analysis of color/simultaneous contrast, it appears as though al-Haytham anticipated the work on this subject by Helmholtz and Hering (Kingdom, 1997).

DARK AND LIGHT ADAPTATION

In his Book III, Chapter 5 al-Haytham deals with 'on the ways in which sight errs in pure sensation'. In this chapter, after a discussion of errors of recognition, he gives examples of errors depending upon the visible properties when the relevant properties fall outside what he calls the 'moderate' range. In the case of luminosity, we can interpret these to be between low scotopic and high photopic levels.

He is somewhat hazy in his description of adaptation phenomena. He starts out with the statement:

> ...errors in [perceiving] light qua light can occur only in regard to the light's strengths and weaknesses, for that which sight perceives of light qua light is nothing but luminosity (p. 262).

He recognizes that the light level is important for seeing color. He states:

> Sight errs in pure sensation when the light in the seen object falls outside the moderate range. This happens for example, when again a variously colored object in which the colors are strong and closely similar is viewed in the light of a small flame. Sight will perceive such an object as being of a single dark color....For sight can perceive the colors of such an object as they are in the presence of a strong light...sight errs in pure sensation when light falls outside the range of moderation (pp. 264–265).

From examples of failure of sight of pure sensation, he then identifies the way to avoid them, namely adaptation, that is, time. Of course, he had no knowledge of photochemistry of the visual pigments. However, he correctly identifies that it takes a certain amount of time:

> sight errs in pure sensation when the duration [of perception] falls outside the range of moderateness.

He gives an example of a colored object in a dark place and says that if you glance briefly at the object you will not perceive the color. However,

> if, however, the eye remains fixed before such an object for an extended interval of time, sight will perceive the various colors provided the place is not too dark (p. 266).

The failure to recognize colors is due to the fact

> "...that a very faint light does not have an effect on the eye" (p. 266) and in order to see it "faint light can produce a sensible effect in the eye only after a fairly long interval of time, for faint light has little strength and is of no effect" (p. 266).

This last statement about time can be interpreted as temporal summation or Bloch's law qualitatively (Gorea, 2015).

In addition,

> when the eye looks at a strong light for some length of time, then turns to look at ta white or light colored object ...light will find it dark (p. 266).

He then makes the claim that in cases of pathology,

> ...if the eye suffers from an illness that dims its sight it will perceive the colors of the object as dark and turbid, that is, as other than they are and therefore will mistake the colors of these objects.

DIRECTIONAL SENSITIVITY

The eye is directionally sensitive – that is, light rays coming in through the center of the pupil are more effective in eliciting a percept/response than rays that are incident more obliquely. This phenomenon was discovered in 1933 and is called the Stiles Crawford effect of the first kind (Stiles and Crawford, 1933). This effect is due to the fact that retinal photoreceptors are oriented towards the center of the exit pupil of the eye (image of the iris formed by the lens) and they act as optical waveguides efficiently trapping the incident light and guiding it to the sites of absorption by the photopigments (Lakshminarayanan and Enoch, 2010). Where does al-Haytham fit in this?

In his explanation of the intromission theory, al-Haytham regards the surface of the object to be made up of a number of points (or patches) which when illuminated send out rays of light in all directions. Some of these rays enter the eye through the pupil and give rise to the visual process and finally the percept of the object. Russell (1996) has a very good description of the process as theorised by al Hazen. Al Hazen does a punctiform analysis (Section 'Anatomy of Eye and Schematic Eye Models'). Since he was the discoverer of the camera obscura, he knew that the pupil was too large to allow the eye to work as a pinhole camera. With large pupils, the pattern of rays would be degraded and the image would be lost. The pupil would not cut out the multiple rays that reach it simultaneously from each point on the surface of the object. He reasoned that the visual rays would intermingle and confusion would arise. He solved this problem by using his knowledge of refraction. Sabra points out that the solution he found was derived from mechanics (Sabra, 1964). Al-Haytham applied mechanical analogies to the effect of light rays on the eye and imposed a condition that of all the visual rays emanating from a single point of a body's surface, only the ones which hit the cornea perpendicularly, and hence not bent by refraction, contribute to the formation of the image in the interior of the eye.

He writes:

> if the eye senses the visible objects through forms that come to it from the surfaces of those objects, then the eye will perceive only those forms of the objects that reach it along the straight lines whose extremities meet at the center of the eye.

From geometry (and his schematic eye; Figure 6.2), it can be seen that these specific rectilinear rays are those that are perpendicular to the cornea (i.e., along a radius) and

> ...if sight senses the visible objects through the forms that reach it from the colors and lights of those objects, and [it senses them] distinctly, then the center of the eye's surface and the center of the surface of the crystalline lens must be one common point and the eye should not perceive any of the forms of the visible objects except though the straight lines whose extremities meet at this center alone (pp. 75–76).

He justifies this by using examples in nature (p. 76).

In fact, al-Haytham believed that refracted rays must lose their power to stimulate the sensitive organ. He is rather vague in explaining how the crystalline only sensed these central rays and not the oblique rays. The Stiles Crawford effect essentially states that light rays normal to the retinal surface are more likely to be detected and this can be inferred from al-Haytham's analysis. The real explanation is given only in terms of waveguide theory, which of course al Hazen could not even imagine. It should also be pointed out that in his use of mechanical analogies to explain optics, he was anticipating the formulations of optics by Lagrangian and Hamiltonian methods which are derived from classical mechanics (Lakshminarayanan et al., 2001).

EYE MOVEMENTS

It appears as though Ibn al-Haytham was the first to recognize the importance of eye movements for perception. As we know, without eye movements there is no perception of the outside world because of neural adaptation.

In his studies of eye movements, he starts out with the observation that the anatomy is such that:

> ...when the eye moves it moves as a whole ...for the eyeball is enclosed in this concavity and it moves as a whole inside it (p. 57).

In this statement, the concavity refers to the orbit.

From his analysis, he states that the various tunics of the eye keep it rigidly in place and the only motion proper to the eye is rotational motion up-or-down or side-to-side. In this, he seems to anticipate Listing's law (named after the German mathematician Johann Benedict Listing, 1808–1882). Consider this: the eye rotates with three degrees of freedom. This means that the eye can rotate about: (1) a vertical axis to generate horizontal eye movements (abduction and adduction), (2) a horizontal axis to generate vertical eye movements (elevation and depression), and (3) the line of sight to generate torsional eye movements (excyclotorsion and intorsion). In theory, the eye could assume an infinite number of torsional positions for any gaze direction. If there are an infinite number of possible torsional positions for each gaze direction, does the eye adopt one or multiple torsional position(s) for a particular gaze direction? The answer to this question is provided by Listing's law. It states that, when the head is fixed, there is an eye position called the primary position, and that the eye assumes only those orientations that can be reached from the primary position by a single rotation about an axis in a plane called Listing's plane. This plane is orthogonal to the line of sight when the eye is in the primary position (Wong, 2004; Tweed et al., 1990). However, al-Haytham has not considered torsional motions.

He gives an aetiological reason for why the eyeball is round and discusses foveation (i.e., view of object by the fovea, the area of clear, sharp vision) and how it is achieved by saccades.

> the eyeball is rounded because roundedness is the most balanced of shapes and also the easiest to move. But the eye needs to move and to move quickly in order to face – though movement – many visible objects at the same time and from one position of the person, and in order to confront – though movement – all parts of the object with the middle of the seeing organ and thereby perceive it clearly and uniformly. For sensation through the middle of the sentient organ is clearer than sensation through the remaining parts…but the quick movement of the eye and the need of the eye to move quickly is in order that it may – through quick movement – contemplate all parts of the visible object and all visible objects facing it in the least time (pp. 103–104).

As we know, saccades are ballistic movements of the eyes made to reposition our gaze and to foveate a target. They can be voluntary, but normally they are automatic and are unnoticed. In addition, during the movement, the resulting image motion and the repositioning of the gaze are not noticed. The first to document this was al-Haytham:

> For if the eye moves in front of visible objects while they are being contemplated, the form of every one of the objects facing the eye … will move on the eye as the latter moves. But sight has become accustomed to the motion of the objects' forms on its surface when the objects are stationary, and therefore does not judge the object to be in motion (p. 194).

It is also interesting to note that al-Haytham seems to have 'discovered' the fact that saccadic eye movements are time optimal (Clark and Stark, 1975; Bahill and Stark, 1978)!

To sum up, it appears as though al-Haytham was the first to recognize the important role played by eye movements in visual perception. It is only in the past few decades that we have realised that there is no perception without such movements and that such movements are crucial to our percept of the external visual world (e.g., Yarbus, 1967; Tatler et al., 2010). The interaction between eye movements and visual perception is optimised for the active sampling of information across the visual field by eye movements and for the calibration of different parts of the visual field. The movements of the eyes and information available to the system are well entangled. These two processes interact to enable an optimal perception of the world (Gegenfurtner, 2016 for an interesting review).

THE HOROPTER AND BINOCULAR VISION

Al-Haytham investigated how humans form a single fused image of the visual world from the images formed in the two eyes which look at the world from different viewpoints because of disparity. A good comprehensive review of his contributions to binocular vision can be found in Reynaud (2003). Al-Haytham, in his studies, uses the analogy of light transmission in a beam chamber to explain how the images combine in the optic chiasm:

> A beholder, however perceives visible objects with two eyes. But if vision is brought about through the form that occurs in the eye and the beholder perceives the objects with two eyes, then the forms of visible objects occur in both eyes ....the beholder perceives each object in most cases as one. The reason for this is that the single object's two forms that occur in the eyes ...come together when they reach the common nerve and coincide with one another and become one form. And from the from thus united ...the last sentient perceives the form of the object (p. 85).

Al-Haytham describes a point-to-point punctate analysis of projection and superposition of the two images coming from the two eyes. The separate points were combined in the 'common' nerve and, if they made an exact match, were fused into a single image. It should also be emphasised that al-Haytham supported his conjecture of image fusion or matching based on his knowledge of the role of eye movements for binocular integration.

He then goes on to discuss diplopia and shows that it is proof that the final fusion of the two images occurs elsewhere.

> ... the fact that a single visible object is in some cases perceived as one and in others as two, while in both cases it has two forms in the eyes then proves that there exists besides the eye a sentient in which a single form is produced for a single object....thus sensation is effected only that sentient and not by the eye alone" (p. 86).

He reasons that if the two images were spatially misaligned, diplopia would result and, owing to the topologically non-overlapping images, no fusion would occur in the chiasma, resulting in diplopia. It should be pointed out that he does not seem to have considered the disparity as the basis for stereopsis.

He also describes vision with objects fore and aft of a fixation point (both object and fixation point in the median plane). He starts out by saying that for objects lying far from the fixation point,

> the form of an object that lies aside from the two axes and far from their meeting point of these axes will be undefined and indistinct as long as the object remains distant from

the meeting point, but the form of the object will become distinct when the axes move and meet on it....while the eyes perceive another object closer to or farther from them than that on which the two axes meet, while at the same time being located between those axes then that object will be differently situated with reference to the eyes in respect of direction. Because if it lies between the two axes, then it will be to the right of one of them and to the left of the other; the rays drawn to it from one eye will be to the right of the axis and those drawn to it from the other eye to the left of the axis; therefore its position relative to the eyes will differ in respect to direction. The forms of such an object will occur in differently situated places in the eyes; the two forms produced by it in the eyes will proceed to two different places in the cavity of the common nerve... and there will be two non-coincident forms (p. 237).

In Book III, al-Haytham describes binocular vision experiments with a board (Figure 6.3; p. 238) and derives a set of conclusions based on these results. With this simple set-up, he was able to define both corresponding visual lines and crossed and uncrossed visual directions. It should be noted that Ptolemy (Lejeune, 1956, pp. 104–105) had used a similar board and stated that singleness of vision occurred when the two visual directions corresponded and introduced the concept of correspondence in binocular vision. He also defined the 'visual' axis and the 'common' axis. Al-Haytham made a similar board on which he placed different colored wax cylinders. Sabra (1966) has translated a manuscript by al-Haytham in which he critiqued Ptolemy's findings using the board and the conclusions he drew from them. However, even though al-Haytham had pioneering findings on corresponding points and binocular disparity, he did not connect the geometry of binocular vision with perception of depth. Al-Haytham called the line falling along the bisector of the visual axes the common axes, *a la* Ptolemy, and the point where the images from the

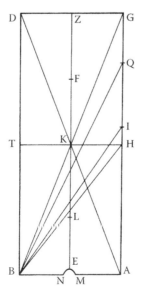

FIGURE 6.3   Board used for binocular vision experiments (see text for details).

two visual axes converged as the center (p. 235). The common axis is now termed the cyclopean axis. His description of the common axis is surprisingly modern:

> let us imagine the line that joins the centers of the apertures to be bisected and imagine a line drawn to the bisecting point from the middle point in the cavity of the common nerve at which the lines extending though the nerve's cavities meet – this line will be perpendicular to the line that joins the centers of the apertures. Imagine this line to extend in the outward direction facing the eyes – it will be fixed in one unvarying position because the point at the middle of the cavity of the common nerve in which the lines extending through the middle of the cavities of the nerves meet is one and unchanging. But the point that bisects the line joining the centers of the apertures is one and unchanging. Therefore the position of the straight line that passes through these two points is one and unchanging. Let us call this line the "common axis" (pp. 232–233).

Al-Haytham also describes basic facts about cyclopean visual direction which are now called Hering's law of visual directions. Referring to Figure 6.3, if we view the board extending from the bridge of the nose (Bridge of the nose is at point E; the eyes at points A and B) and draw a line EZ down the center of the board, it will appear as though it were two lines intersecting at fixation point K. Two other lines AD and BG extending diagonally and intersecting at the fixation point will appear as four lines, with the center two appearing close together at the median plane of the head. Al-Haytham's explanation for this is that these two lines appear close together near the center because they lie close to the visual axis and hence a common visual direction in the fused image. This visual direction axis lies along the 'common' axis that bisects the angle between the visual axes and intersects the center:

> The reason why two of the four appear closer together is this: when the two visual axes meet at the middle object, then each of these two diameters will be perceived by the eye next to it thorough rays that are very close to the visual axis; thus their two forms (images) will be very close to the center within the common nerve (the chiasma) and their point of intersection will be at the center itself, and thus the diameters will appear very close to the middle (the median plane) (p. 242).

and

> ..an object will be seen double if the rays that meet on it have the same direction but differ greatly with respect to their distance from the two axes; that an object will be seen double if it is perceived through rays of different directions through their distances from the two axes may be equal (pp. 241–242).

From this, it is easy to discuss stereoscopic vision:

> The axes of the eyes may meet on an object while the eyes perceive another object differently situated in direction with respect to the eyes. This happens when the object lies closer to the eyes than that on which the axes meet while being situated between the axes, or when it is farther off from the object on which the axes meet while also being situated between the axes if imagined to be extended beyond the meeting point

provided that the object on which the axes meet does not obscure the farther object or obscures (only) part of it (p. 232).

Al-Haytham also describes crossed and uncrossed visual directions:

if the two axes of the eyes meet at a point on the surface of the object facing the eyes, then that surface will be the common base of the radial cones formed between the centers of the eyes and the object; the point in which both axes meet will have the same position with respect to both eyes because it will be opposite the middles of both eyes and the axes between it and the two eyes will be perpendicular to the surfaces of those eyes at their middles. As far the reminder of the object's surface, there will exist between every point in it and the centers of both eyes two lines similarly situated in direction with respect to the two axes (p. 229),

followed by:

if, however, the two axes meet on a visible object, while the eyes perceive another object closer to or farther from them....then, it will be to the right of one of them and to the left of the other; the rays drawn to it from one eye will be on the right of the axis and those drawn to it from the other eye to the left of the axis; therefore its position relative to the eyes will differ in respect to direction (p. 237).

Both in Books I and III (Chapter 2), al-Haytham describes the concepts of corresponding points. Ptolemy only described corresponding points in terms of visual lines. Using corresponding points, he was able to show how we see a single image when the two ocular images fall on corresponding points and how diplopia occurs when they fall on non-corresponding points.

Al-Haytham gives instructions to a subject (the reader) to view the board shown in Figure 6.3 extending horizontally from the bridge of the nose. Then he places small objects (two waxed cylinders) at K.F,L,I and Q. When the eyes converge on K, the line EZ would appear like a cross passing through K and the lines BG and AD would appear as four lines with the two center lines superimposed down the midline. Objects F and L would now be seen double, one on either side of the midline. Q would produce double images on both sides of the midline while object I would double images that are too near to be seen as two. Objects T and H would appear single, but objects more eccentric on the same horizontal plane would appear double. The object at I would not be seen as double if it is not too far from the plane TKH on which the eyes are converged while the object at Q far away from the plane of convergence would appear diplopic (pp. 237–241).

From this, al-Haytham then states,

the two points in the eye's surfaces that lie on the axes are similarly situated with respect to the cavity of the common nerve, and the same points are similarly situated with reference to every point on the common nerve; therefore the two points in the surfaces of the eyes that lie on the two axes will be perfectly similarly and equally situated relative to the point on the common axis at the middle of the common nerve where the lines drawn from the centers of the apertures meet. Thus when the two forms that occur in the two points where the surfaces of the eyes intersect the axes reach the common nerve, they will both occur in the point on the common axis that lies in the middle of the cavity of the common nerve where the lines meet thus becoming single form (p. 233).

Also, more specifically,

..every pair of forms at two similarly situated points in the eyes will proceed to a single point among those surrounding the point on the common axis; and thus the two forms of the whole object will coincide with one another and become one and the object will be perceived single, This then is the manner in which the two forms produced in the sight for a single object similarly situated in relation to the eyes become one, and the manner in which the sentient perceives a single object as one though two forms of it are produced in the eyes (p. 234).

Having described the corresponding points, al-Haytham then goes on to show that the lines from the object need not fall exactly on these 'corresponding' points, but can be off by a little:

If, …the object lies outside the common axis but is not excessively far from it, then the two forms produced by it in the eyes will not greatly differ, and therefore the form of it produced in the common nerve will not be double. If the object lies outside the common axis and is excessively far from it, while the axes of the eyes meet at a point in it,……. .the object's form will not be distinct but confused…if the object is of a small size and approximately equal dimensions, and it lies on the common axis or close to it, the from produced by it ..will be one and also distinct (pp. 235–236).

and further:

if the object lies on one of the axes and outside the other it will produce in the cavity of the common nerve two non-coincidental forms one in the center and the other displaced from the center (p. 237).

In the above statements, al-Haytham is basically describing fusional areas, the region over which disparity is tolerated without producing diplopia, a concept that is now attributed to Panum (1858).

Al-Haytham then goes on to study diplopia in detail and provides evidence that the locus of fused images for a given viewing distance does not lie in a frontal plane and describes what is now called the horopter. The modern concept of the horopter (Greek: 'horos' – boundary and 'opter' – observer) is attributed to Franciscus Aguilonius (1613) who describes it in his Book II of his six volume opus, *Opticorum Libri Sex*. As Howard (1996) notes, al-Haytham described the concept of corresponding points and the horopter in its modern definition as the locus of fused images five centuries before Aguilonius was born.

Al-Haytham once again bases his arguments on observations with his board. Referring to Figure 6.3, he first observes that objects closer to or farther than fixation point (L and F) appear double and on opposite sides of the fixated object when they are between the visual axis and on the same side of the fixated object when they are outside the visual axis (Point Q). He also points out that objects H and T in the same frontal plane as the fixation point will also appear single if they are not too far from the fixation point but will be seen double if they are far away. He writes:

all familiar objects face both eyes when they are regarded by both eyes; thus the two visual axes always meet upon them; the remaining rays that meet on every point in them lie on the same side; and no great difference exists between their distances from the two axes, therefore every familiar object is seen single by both eyes. It is only rarely that objects are seen double for an object is seen double only if it differs greatly relative to the two eyes with regard to direction or with regard to distance or both (p. 242).

Al-Haytham uses analytic geometry to show that the angles between the median plane and the eccentric object are not the same for both eyes (what we now call longitudinal angles; Ogle, 1965). However, he defines the horizontal locus of fused images as those for which angles of binocular subtense are equal, which we now call the horizontal horopter. From this, he shows that the locus of fused images for a given fixation distance does not lie in a frontal plane.

Once again, using experiments and logic with his board, he gives evidence that an object vertically situated below or above a fixation point in the midline will be seen as single because the distances from the two eyes remain the same. In this, he foresees the vertical horopter as the locus for which the distance to the two eyes is the same (Cogan, 1979; Joseph and Lakshminarayanan, 2001). This concept of the vertical horopter is attributed to Helmholtz. It should be pointed out that al-Haytham (like Ptolemy) assumed that the horopter, described as the plane of the visual field, was flat. However, we now know (since the time of the Vieth-Muller circle in the first decades of the nineteenth century) that the horopter is curved. However, to his credit, the description of the horopter as a set of points as a single vision was recognized by al-Haytham. He also used the singleness as the criterion for the horopter – one of the criteria that are commonly used today (Ogle, 1965).

LAW OF EQUAL INNERVATION

Hering's law of equal innervation is used to explain conjugate eye movements, and essentially, it refers to the fact that the extraocular muscles for each eye are innervated equally (Hering, 1977). Aristotle and Ptolemy seem to have believed that there was a single unified source to control eye movements in both eyes (Howard and Wade, 1996). It seems as though it was Descartes in 1626 who conceived the model for the controller, a muscular mechanism for the movement of the eyeballs, and of reciprocal innervation as the cause of the agonist/antagonist muscle pair in each eye (Ciuffreda and Stark, 1975).

Al-Haytham carefully observed eye movements in subjects and stated (Chapter 2 of Book III):

we say, then, when a beholder looks at a visible object, each of his eyes will regard the object; when he gazes at the object, both of his eyes will equally and similarly gaze at it. ...when sight moves over the object in order to contemplate it, both eyes will move over it and contemplate it.

He then continues on to a long geometric analysis (using the motion of the axes he had defined; Figure 6.2). This analysis is in the context of binocular vision. As a precursor, he writes,

when the beholder fixes his sight on an object, the axes of both eyes will converge on the object, meeting at a point on the surface. When he contemplates the object the two axes will together move over the surface of the object and together pass over all it's parts. And in general the two eyes are identical in all their conditions...their actions and affections are severally always identical. When one eye moves for the purpose of vision, the other moves for the same purpose and with the same motion; and when one of them comes to a rest, the other [likewise] is at rest.

This paragraph concludes with:

Thus it is not possible that one eye should move for the purpose of seeing while the other remains motionless, nor that one eye should strain to look at an object without the other straining to look at the same object, unless some obstacle or cover or some other accident intervened, thus hindering one of the eyes from participating in the act performed by the other. When both eyes are observed as they perceive visible objects, and their actions and movements are examined, their respective actions and movements will be found to be always identical ...when both eyes are observed as they perceive visible objects and their actions and movements are examined, their respective actions and movements will be found to be identical (p. 229).

The above statements imply that the extraocular muscles be equally innervated in order to achieve the conclusions of al-Haytham. Hering's important additional contribution was the statement that the other eye, even if closed or blind, follows exactly the trajectory of the seeing eye – a point also made by the Scottish philosopher Thomas Reid 100 years earlier. However, in his description of all the other features of conjugacy al-Haytham was correct. In fact, Heller (1988) in his history of eye movement research credits al-Haytham for this discovery.

THE MOON ILLUSION

The moon illusion is one of the oldest visual illusions recorded by human cultures in human history (Ross and Plug, 2002; these authors show that a cuneiform inscription on a clay tablet from Nineveh dating back to the seventh century BCE describes the illusion. This illusion has been studied by Aristotle, Ptolemy and later Islamic scholars and extends to Leonardo da Vinci, Descartes, George Berkeley and, in the twentieth century, by perceptual explanations based on methods of experimental psychology by Boring (1943), as well as Rock and Kaufman (1962), Kaufman and Rock, (1962) and Kaufman and Kaufman (2000). The book by Hershenson (1989) has a number of chapters dealing with various aspects of the illusion.

What is the moon illusion? It is simply a perceptual phenomenon that the moon at the horizon may appear many times the diameter of the elevated moon. Attention should be called to the fact that irrespective of the position of the moon in the sky, the angular size (the visual angle subtended at the eye) remains the same whether the moon is at the horizon or elevated. This is a very powerful real-world illusion. A rather comprehensive review of the history and theoretical analyses is given by Nanavati (2009).

It is also interesting to note that the explanation of the illusion at various time periods provides support to the paradigm model of Thomas Kuhn. That is, explanation of natural phenomena come from a 'model from which spring particular coherent traditions of scientific research' (Kuhn, 1970, p. 10).

Aristotle (ca. 300 BCE) related these size changes to the effects of mist (Ross, 1931, p. 373b):

> ...promontories in the sea loom when there is a southeast wind, and everything seems bigger, and in a mist, too things seem bigger.

Ptolemy (ca. 50; 1952) gave a similar argument in his *Almagest* even though he understood that objects in a rarer medium appear smaller. However, in his comments on the illusion in his other two books, namely *Optica* and *Planetary Hypotheses*, he offers explanations of other aspects of the illusion (Sabra, 1987). The passage in *The Almagest* draws on the analogy of the 'apparent enlargement of objects in water, which increases with the depth of the immersion.' (Ross and Plug, 2002, p. 7). However, in his *Optica*, he points out that this refractive explanation is not correct since the moon would be in the rarer medium (the sky), and hence the moon should look smaller. This explanation is physical rather than perceptual. Ptolemy rectified this in his latter book *Optica*; in the third book of this treatise, he gives a psychological account based on apparent distance. He writes:

> For generally, just as the visual ray, when it strikes visible objects in (circumstances) other than what is natural and familiar to it, senses all their differences less, so also its sensation of the distances it perceives (in those circumstances) is less. And this is seen to be the reason why of the celestial objects that subtend equal angles between the visual rays, those near the point above our head look smaller, whereas those near the horizon are seen in a different manner and in accordance with what is customary. But objects high above are seen as small because of the extraordinary circumstances and the difficulty (involved) in the act (of seeing) (Sabra, 1987, p. 225).

There are some inconsistencies in these two ideas (Sabra, 198). In the *Optica*, Ptolemy applies axiom 1:2 of Euclid and develops what is known now as the law of convergence for visual perception. That is, the visual axes have to converge towards the object of regard in order to avoid double vision (diplopia; however, he did not associate it with binocular vision!). It is also interesting to note that he bases this law on the assumption of an egocentric reference frame. He then attributes a psychic element to his theory of vision and also enunciates the size-distance invariance (McDonald and O'Hara, 1964).

In his Astronomy book, *Planetary Hypotheses*, Ptolemy writes that the moon illusion is

> an error that occurs to sight on account of the difference in perspective (unique) to large distance (Sabra, 1987, p. 160).

That is, according to Ptolemy, the illusion is due to the fact that the distances are large beyond the normal range of convergence.

Al-Haytham accepted both ideas (refraction as well as psychic interpretation). In his *Commentary on the Almagest*, he provides geometric proof through which the size of the moon could be magnified by refraction, thus validating Aristotle. He shows that the moon's light, when refracted by the atmosphere, could be made to look much bigger using mathematical proof. Al-Haytham, then in his next book *Solution of Difficulties in the Almagest Which a Certain Scholar has Raised*, writes:

> Now if someone believes that the magnitude of an object is perceived according to the magnitude of the angle alone then he will doubt Ptolemy's statement (in the Almagest) that the object increases in size as it sinks deeper. But the statement is subject to doubt only if the doubter relies exclusively on the angle. Such a reliance however, would be a mistake; for Ptolemy has shown in the Optics, in the course of his discussion of size that the size is not perceived according to the angle alone but according to the magnitude of the angle and of the (object's) distance and according to whether the object is inclined or frontally situated. Thus, Ptolemy says in the second of his Propositions on size that of two unequally distant objects and perceived through the same angle, the nearer looks larger than the farther object, but is either seen as smaller (than the farther object) (when a sensible interval exists between them) or equal to it (when the difference between their distances (from the eye) is insensible. Thus if size were perceptible by means of the angle alone, then two objects seen through the same angle, or through two equal angles would always appear to be equal – which is not the case. Moreover, we have shown in our book on Optics, that size is perceptible by (comparing) the angle to the distance and also in accordance with the angle alone (Sabra, 1987, p. 233).

Al-Haytham provides his most complete psychological statement on the moon illusion in Book III (On the Errors of Direct Vision), Chapter 7 (Errors of sight in inference when the illumination in the visible object falls outside the moderate range) of his *Kitab al-Manazir*. This analysis is preceded by a discussion of refraction and viewing of heavenly bodies. He begins by explaining the conditions under which sight perceives size:

> What sight perceives regarding the difference in the size of the stars at different positions in the sky is one of the errors of sight. It is one of the constant and permanent errors because its cause is constant and permanent. The explanation of this is (as follows): Sight perceives the surface of the heavens that faces the eye as flat and thus fails to perceive its concavity (VL: the so-called "vault of heaven") and the equality of the distances (of points of it) from the eye. . Now sight perceives those parts of the sky near the horizon to be farther away than parts near the middle of the sky; and there is no great discrepancy between the angles subtended at eye center by a given star from any region of the sky; and sight perceives the size of an object by comparing the angle subtended by the object at the eye-center to the distance of that object from the eye;

therefore it perceives the size of the star (or the interval between two stars) at or near the horizon from comparing its angle to a large distance and perceives the size of that star (or interval) at or near the middle of the sky from comparing its angle (which is equal or close to the former angle) to a small distance (p. 241).

He also says (p. 237),

sight perceives size from the magnitude of the angles subtended at the center of the eye and from the magnitudes of the distances of the visible objects and from comparing the magnitudes of the angles to distance.

Al-Haytham also provides an objects-aerial perspective interpretation: that is, we consider Figure 6.2 in relation to ground cues. Since the horizon moon finds itself in a terrestrial frame of reference, the moon, based on its brightness, is placed in relation to other objects. Just as in an aerial perspective where a brighter object appears farther, the mind makes an unconscious inference against a flat sky. He recognizes that, even though his mathematics on subtended angles could be used to explain the appearance of stars, it cannot explain the horizon moon. He also does some "parameter fitting" – that is, he adjusts for variations in atmospheric refraction. He states that it is a tertiary factor accounting for the variation in the size. He concludes emphatically that the moon illusion is a psychological construct (Hershenson, 1989, p. 22) but does bring in the effect of physics – and incorporates light and classifies it as an error of illumination.

Whatever the shortcomings, al-Haytham's conclusion that the size of the moon is perceived (based upon perceived distance and visual angle) is the underlying central idea in all other explanations of the moon illusion (Kaufman and Kaufman, 2000).

MOTION PERCEPTION

The percept of motion is studied in Volume II of the Kitab in Section 'Visual Psychophysics and al-Haytham' (pp. 193–196). Al-Haytham gives three general ways by which we see motion:

...by comparing the moving object with other objects, or with a single object or with the eye itself...Now sight perceives the motion of an object by perceiving the object in two different places or positions. And an object's position can vary only in time...but a certain duration must exist between any two different moments and therefore sight can perceive motion only in time (p. 193).

As seen in the above quote, al-Haytham makes a valid point that motion can be detected only with reference to one or more objects or with respect to the observer. It is also important to note that by saying motion can be detected by the eye as a reference, he underscores the fact that eye movements are involved in motion perception.

Ibn al-Haytham gave concrete meaning to the temporal threshold:

For sight perceived motion only by perceiving the object in two different places one after the other or in two different positions one after the other...For if sight perceives the moving object in the second position without at that moment perceiving it in the first position where it was formerly perceived then it will sense the difference between the two moments. And if it senses that difference, then it will sense the time between them. That being the case, the time in which sight perceives the motion must be sensible.... sight can perceive motion only in time, for motion and every part of motion must take place in time (p. 195).

Al-Haytham also deals with perception of motion when the eye is moving. He says:

There is a difference in condition between the variation in position that happens to that object on account of its own motion and the variation in position that happens to it on account of the eye's motion. Thus when sight senses the varying position of the object, and senses that the variation in object's position is not due to the eye's motion, it will sense the object's motion... (p. 194).

The above statement is the idea of sensations of innervation ascribed to Helmholtz and Wundt (Ross and Bischof, 1981) and the reafference principle. Helmholtz and Wundt believed that the innervation of efferent tracts in voluntary action establishes sensations that arise wholly within the brain and not by way of afferent fibres from the activated muscles. The reafference principle offers an explanation for the fact that stimuli caused by self-initiated motions (e.g., voluntary eye movements) are not perceived (Holst and Mittelstaedt, 1950).

Lastly, he talks about induced visual motion. In this illusion, a stationary or a moving object will appear to move or to move differently because of other moving objects nearby in the visual field. It is interpreted in terms of the change in the location of an object due to the movement in the space around it (Duncker, 1929). A good example of this is the fact that the moon on a cloudy, windy night appears to be racing through the sky opposite to the direction of the clouds, though the moon is essentially stationary in the sky and only appears to be moving due to the movement of the clouds.

In Book III, Chapter 4 he states the problem:

An example of the errors that may occur in inference and discernment at the time of perception is provided by sight's perception of the movement of the moon when a thin, discontinuous cloud of variable form moves fast before the moon's surface. Sight will erroneously see the moving as moving rapidly (p. 260).

And his solution to this question is:

for when a transparent body moves in front of objects placed on the ground near the eye, they are not perceived to be in motion, provided the remaining conditions of veridical perception for these objects are within the range of moderateness. This may be experienced in bodies on the surface of the earth, namely those immersed in clear running water. For, observing an object immersed in clear running water, sight will not perceive that object to be in motion, though it perceives the moon as moving behind a cloud (p. 261).

And correctly gives the analysis in terms of:

The inference on account of which this error occurs involves correlating the moon with parts of the cloud (p. 261).

Induced movement was reported by Ptolemy but al-Haytham has once again gone beyond Ptolemy to discuss this as a psychological error of inference and hence a visual illusion.

PATTERN CLASSIFICATION AND OBJECT RECOGNITION

Object or pattern classification is a major area in computer vision and artificial intelligence. In particular, the so-called Nearest Neighbour (NN) technique is often used, which is a simple and efficient method in pattern recognition, object recognition, event recognition, etc. The NN rule assigns the category of unknown data points based on the nearest point (or neighbour) of a set of previously classified points. This decision rule was first formulated by Cover and Hart (1967). Their method is called the k-NN and is based on the calculations of the value of k, which specifies how many nearest neighbours are to be considered to define the class of a sample data point. The various nearest neighbour methods are discussed by Bhatia and Vandana (2010). Another major area of interest currently is the field of neural networks and deep learning. A neural network is loosely based on biological neural organisation. In a neural network, a large number of initial data, say images of faces, is used to 'train' the network and the network develops a system which can learn from these examples, and then when an unknown face is presented, it compares the test face with the rules it has learnt from the training set (Schmidhuber, 2014 for a good review) and categorises the image face. A good review of deep learning methods in vision science and opthalmology is given in Leopold et al. (2017).

Al-Haytham gives a very clear formulation of NN classification and neural networks (Pellilo, 2014). He gives importance to the role of memory and inference and argues the importance of similarity in object recognition and classification. This observation of the importance of similarity anticipates the Gestaltian idea (Koffka, 1935). These arguments are given in Book II of *Kitab al-Manazir*.

In Chapter 3 of Book II, he first says:

…among the properties that are perceptible by the sense of sight, some are perceived by pure sensation, others by recognition and others still by a judgmental inference that exceeds the inference of recognition (p. 130).

What is crucial for pattern classification is in the 'recognition' aspect. He goes on to further write:

"sight recognizes visible objects and it perceives many of them and of the visible properties by recognition" (Sabra, p. 128) and states emphatically, "It is only by recognition that sight perceives what a visible object is. And recognition is not perception by pure sensation since sight does not recognize all that it has seen earlier" (p. 128).

What is the role of memory in pattern recognition/classification? He has an answer to this question also:

If recognition without remembering is not possible, then recognition is not perception by pure sensation, but rather it is perception through a kind of inference. For recognition is the similarity of two forms and he goes on to state that it is a matter of seeing the similarity between what is seen and the remembered image and further states:

> It is for this reason that recognition cannot take place without remembering, because if the first form were not present to the memory, sight would not perceive the similarity of the two forms, nor would it recognize the objects (p. 129).

This perception is inferred by comparison –

> Now perception of likeness is perception by inference because it occurs by comparing one of the forms with the other. Recognition is therefore due to a kind of inference

and

> but this inference is distinct from all [other] inferences .For recognition does not occur as a result of inspecting all properties in the form, but rather through [perception of] signs. For when sight perceives one of the properties in the form while remembering the first form, it recognizes the form" but importantly "that is not so with all that is perceived by inference for many such things are perceived only after inspecting all or many of the properties of the individual object perceived by inference (p. 129).

We recall here the modern, computational idea of vision, namely that the vision is a mathematical process in which certain operations are performed on the retinal image and 'That we see the world as well as we do is something of a miracle. What seems so direct and effortless turns out, on close consideration, to involve many rapid and complex processes the details of which we are only beginning to glimpse' (Crick et al., 1980).

How do we then categorise objects? Al-Haytham now brings in the concept of 'universal forms' – which are nothing but categories in modern terminology (or primitives in computer vision) and describes how the categorisation is done by the 'Soul'. The universal forms are generated and

> sight will ascertain the form of a visible object only by scrutinizing all parts of the object and contemplating every feature that may be apparent in it (p. 209).

This idea of primitives or a template for various objects is described further:

> Now for the universal form which are produced by in the soul for the species of visible objects and which takes shape in the imagination. To every species of visible objects belong an appearance and shape which are the same for all individuals of that species, while the individuals differ in respect of particular properties which are also visible. Color [for example] may be the same in all individuals of one species. Now appearance, shape, color and all properties which constitute the appearance of every individual of a certain species is a universal form of that species (p. 213).

In an analogy to neural network training, he further states

> And when the universal form has been repeatedly presented to the soul, it will be fixed and established in it. And the difference between the particular forms that accompany the universal forms as they are repeatedly presented, the soul will perceive the form that is identical for all individuals of the species is a universal form of that species. In this way, then, the universal forms which sight perceives of the species of visible objects are produced in the soul and in the imagination (pp. 213–214).

In addition,

> the perceived forms of individual visible objects….as they are repeatedly perceived by sight they become more firmly fixed in the soul and the imagination; and visible objects are recognized by the sentient by means of the forms produced in the soul for the species of these objects and their individuals. It is on these forms that the sentient relies in perceiving what the visible objects are, because perception of what they are is due only to recognition and recognition results from comparing the form presently perceived by sight with the form that has been fixed in the soul by the forms of objects already seen, and from likening the presently perceived form to one of the forms in the imagination. Perception of what the object is, therefore, is perception of the similarity between the object's form and one of the forms established in the soul and in the imagination for the species of visible objects (p. 214).

Note that al-Haytham gives an algorithmic view – the 'soul' is trained with known objects which it then categorises and forms templates. Perception of the image is achieved by studying the similarities between the image and the stored template. This template matching is a very common technique in pattern recognition and computer vision (e.g., Perveen et al., 2013). However, al-Haytham emphasises that it is not just a simple geometric template matching, but takes into account all known knowledge and properties of the object. He discusses how the 'soul' categorises a rose from a general flower:

> when sight perceives roses in bloom in some garden it will immediately perceive that these visible objects are roses on account of the particular color of the roses in addition to their being in a garden, before perceiving the round shape of their petals or their arrangement, and before perceiving all properties that constitute the form of roses. And if the roses resemble some other flowers, sight will perceive them in any case to be flowers and not leaves of trees or [other] plants (Sabra, p. 219).

With these categories, when an unfamiliar object is presented, that is, there is no previous memory counterpart, then

> When, therefore, sight perceives an object, the faculty of judgment will look for a similar form in the imagination. If it finds such a form, it will recognize the object and perceive what it is; if not, then it will neither recognize the object nor perceive it's quiddity. However, because of the speed with which the faculty of judgment assimilates the form of the object at the moment of vision, it may err by likening the object to another, different from it, if the object has a property which exists in the other. Then, when it later contemplates the object and ascertains its form, it will liken it to the form truly

similar to it, thus realizing at the second time the error it made in the first assimilation. It is in these ways, then, that the sense of sight perceives what the visible objects are (p. 214).

This is nothing more than what is known as a reject option decision rule. A reject option occurs if a classification does not give an unequivocal decision. In that situation, it may be possible to defer the decision until more information is available or it may be possible to discard the questionable measurement. It has been pointed out that this reject option is particularly easy to incorporate in the k-NN classification rule (Hellman, 1970).

What about the time taken for the matching? al-Haytham gives a discussion that is very similar to the results of the famous experiments reported by Shepard and Metzler (1971) – the more complex an object is, the longer it takes to recognize the object (Shepard and Metzler looked at the time taken to recognize the same object when it was been rotated or reflected), with the conclusion that the brain is doing certain mathematical operations. Al-Haytham states,

Again, the times required for perceiving the species of familiar objects will vary; for some species of such objects might be confused with one another, whereas others might not be so confused. Take for example, the species of a man and of horse: the form of man's species cannot be confused with the form of other animal species; but this is not so with a horse, for a horse resembles many other beasts in general appearance. Now the time in which sight perceives enough of a man's individuality and specific nature to be able to perceive him as a man is not the same as that in which it perceives enough of a horse's individuality and specific nature to be able to perceive it as a horse (p. 218).

SIZE PERCEPTION

Ibn al-Haytham employed throughout Book III of Kitab al-Manazir the term 'error of vision' and attributed various factors, such as distance, position, solidity, shape, size, separation, continuity, number, motion, rest, texture, transparency, opacity, shadow, darkness, beauty, ugliness, similarity and dissimilarity as the cause of these 'illusions'. Using these factors, he analysed a number of visual perceptual phenomena. I will discuss his work on size perception in this section and refer the reader to Sabra (1989) or Smith (2001) for translations and discussions of other topics found in Book III.

According to al-Haytham's theory of vision, the *ultimum sentiens* uses the spatial information present in the cross section of the visual pyramid in order to apprehend the objects of vision other than light and color. However, as the geometry dictates, the arrangement of points within the cross section of the visual pyramid provides direct information about only two dimensions: the third dimension, depth or distance, is not present. Al-Haytham saw that the simple understanding of the geometry of visual stimulation was not sufficient to explain human perception in real life. Consider for example, the phenomenological fact that we typically see circles as circles, even if they are oblique to the line of sight (which would cast an elliptical image on the crystalline), and that we are able to discriminate the sizes of objects, even though this discrimination requires taking distance into account. His analysis included two factors, namely that spatial properties as received at the eye (e.g., an ellipse) differ from objective properties (e.g., a rotated circle), and that we nonetheless see its true form. He concludes that psychological processes must also be invoked in order to explain human perception. While this realisation was also implicit in Ptolemy (Lejeune, 1948, pp. 96–97), al-Haytham went further in developing an account of these psychological processes.

Al-Haytham gives an extensive treatment of the psychology of visual judgements. According to him, the faculty of judgement assigns a particular set of properties to the objects seen (Section 'Unconscious Inference and Perception'). It is in this way that distance, size and shape are perceived.

Distance is apprehended through a judgement of the number of regular-sized intervals composing the continuous ground space between the observer and the distal object, or, if the process occurs frequently, through a judgement by recognition. From the distance to an object together with a visual angle, the size of the object can be apprehended, again through judgement by recognition. Similarly, apprehension of the distance to various points in the field of vision can be used to apprehend the solidity and shape of seen objects. Thus, in general, according to al-Haytham the visual apprehension of the spatial properties of the visual world is made possible by an unnoticed process of judgement. Bauer (1911, p. 54), in his paper on al-Haytham's psychology, writes that in the field of spatial perception al-Haytham 'touched upon a series of the most important psychological problems, and his explanations anticipate in a surprising manner thoughts that were again taken up only in the most recent development of psychology.'

Al-Haytham's work on size and distance, namely perceived size, is proportional to perceived distance and hence anticipated Emmert's law (Emil Emmert, 1844–1911). He realises that the size of the visual angle subtended by an object per se is not sufficient and writes:

> But size is one of the properties perceived by inference and judgment…But in order to perceive size the faculty of judgment cannot be satisfied merely with considering the angle or the magnitude of the portion of the eye that subtends it (p. 175).

and

> For when sight perceives two objects one of which is closer to it than the other, and both subtend the same visual angle at the center of the eye.. and if sight perceives the distance of each of the two objects with certainty, it will always perceive the more remote object to be greater than the nearer (pp. 177–178).

He then goes on to state:

> if perception were dependent on the angle alone, two unequally distant objects subtending the same angle at the eye's center would be seen as equal. But, sight never perceives two such angles as equal, provided that it perceives their distances and makes certain of the magnitude of these distances (p. 178).

He then discusses how the apparent size of an object is constant as it is moved to different distances. Even though the visual angle changes, we still see them as being the same size and this is one of the invariances of vision, namely size constancy (pp. 180–190).

In his discussion of size and distance, al-Haytham treats the role of the continuous ground plane which was first dealt with by JJ Gibson (1950). This idea of the ground plane perception is crucial to human (and robot) mobility (e.g., Se and Brady, 2002). As Gibson states, 'there is literally no such thing as a perception of space without the perception of a continuous background surface'. His 'ground theory' hypothesis suggested that the spatial character of the visual world is given not by the objects in it but by the ground and the horizon. Al-Haytham makes the crucial statement:

> ...sight will not perceive the magnitude of their distances if these distances do not extend along ordered and continuous bodies... (p. 153).

and describes an experiment to prove this (pp. 153–154) and gives a more detailed analysis in pages 179–187. He then discusses the breakdown of size constancy with distance. He uses the term 'moderate' distance and says

> ... the faculty of judgment will perceive the object's size according to the magnitude of the form of its ascertained distance which accompanies the object's form. Only the magnitudes of such objects can be perceived with certainty (p. 190).

Again, in Section 'Visual Psychophysics and al-Haytham' on 'Perception of Size', he deals with other aspects such as how overlap is a cue to relative depth order and also how familiarity is a cue to absolute distance (p. 186). I will end this discussion of size perception with al-Haytham's explanation of percept of distant objects:

> The reason why sight perceives an object at an excessively great distance to be smaller than its real magnitude is that the size of the an object is perceptible only by estimating the object's size by the angle of the cone that surrounds it together with the magnitude of the object's distance (pp. 283–284).

UNCONSCIOUS INFERENCE AND PERCEPTION

As noted earlier, al-Haytham realised that optics is only the first step in perception. There are active processes such as attention, comparison and memory which are required for conscious vision. He realised that a series of logical inferences must occur before the percept. He also imposed the condition that the speed of perception demands that these inferences themselves be imperceptive – that is, unconscious to the observer. As he states:

> Nothing of what is visible, apart from light and color, can be perceived by pure sensation, but only by discernment, inference, and recognition, in addition to sensation.

The idea that unnoticed judgements underlie perception has been in the vision science literature at least since the Optics of Ptolemy and in addition to al-Haytham, Helmholtz (Helmholtz and Southall, 1910/1962) and Rock (1983) have offered different versions of the theory of unconscious inferences. They have used these unconscious inferences in the form of syllogisms, inductive inferences and deductions in predicate logic. Sabra (1978) reviews al-Haytham's theory of unconscious inference, though he does not compare it with modern ideas on the subject (Hatfield, 2002 for a detailed discussion).

He further writes (p. 136):

> it is in the nature of man to judge and to make inferences...without effort of the exercise of thought. Thus familiar inferences of which the premises are evident,... are natural to man.; for at the moment of perceiving their conclusion one is not aware of having perceived them through inference....of things that are perceptible to the sense of sight, some are perceived by pure sensation, some by recognition, and others still by a judgment and an inference that goes beyond recognition...the manner of perceiving the visible particular properties des not in most cases become apparent because of the speed with which they are perceived and the speed of the inference by means of which visible things are in most cases perceived and because the faculty of judgment performs these inferences naturally, employing them not by thinking and exertion but by nature and habit.

It is important to point out that al-Haytham describes three stages of vision/perception. The first stage is the stimulus stage, namely light and color, a low-level process. The second stage is the unconscious inference of complex features, recognition of similarities and differences between objects and finally the third stage of conscious deduction of more complex and sophisticated features. The second stage of the process is nothing but Helmholtz's idea of unconscious inference. This idea of successive stages of the perceptual process can be found in Books II and III of Kitab al-Manazir. In Book II, he writes that the sensory features of light and color are sensed at the recipient surface (i.e., the retina in modern language) and some form of inferential process is required for more sophisticated features such as shape, color, movement, etc., and involves complex interactions and memory. Al-Haytham distinguished between the initial stage of processing and the subsequent higher level processing. Today, we call this initial stage of processing pre-attentive vision; that is, visual perception appears to be functionally divided between an early pre-attentive level of processing in which simple features are coded spatially in parallel and a later stage at which focused attention is required to combine the separate features into coherent objects (Treisman, 1985). He writes

> We now say that sight perceives visible objects in two ways: by glancing and by contemplation. For as soon as sight takes notice of the object, it perceives manifest features...if it contemplates it and inspects all of its forms it will ascertain its form. If it does not contemplate the object and scrutinize all its parts, then it will perceive a non-ascertained part of it.... Glancing perception is non-ascertained perception but contemplative perception is the means by which the forms of visible objects are ascertained (p. 209)

After this statement, al-Haytham goes on to give the differences between sight due to 'glancing' (pre-attentive vision) and 'contemplation' (attentive vision) in Chapter 4 of Book II ('On distinguishing the ways in which sight perceives visible objects'). He then states that features which objects have in common and differences between objects is the basis for visual recognition of objects. He then states that we form essentially a 'database', a representation of all properties of that members of a particular class of objects (he calls these collections of features 'significant traits' or 'signs'), that is,

> Objects which sight has previously perceived and whose form it recognizes and remembers, may be perceived by means of signs, but no so unfamiliar objects (p. 217).
>
> Perception of what the object is, therefore, is perception of similarity between the object's form and one of the forms established and in the soul and in the imagination of the species of visible objects (p. 214). (Section 'Pattern Classification and Object Recognition').

He then argues that familiar objects are perceived faster than unfamiliar objects. That is, if such 'signs' are in the database, we can recognize the object without inspecting all of its features:

> when the objects of vision that are perceptible by inference have been repeatedly perceived by inference and the faculty of judgment has understood their meanings, then that faculty's perception of them when they appear before it after their understanding has been established occurs by recognition without the need to inspect all of their properties; rather the faculty of judgment perceives them through signs. This resulting perception, will, therefore, be among those things that are perceived by recognition without resuming the discernment, comparison and inspection of all properties of those objects (p. 132).

To support this notion, he gives the example that a familiar word can be recognized by noticing just the first and last letters or, as he puts it, 'perceiving the configuration of the totality of the form' (p. 132).

These ideas are nothing but modern ideas of visual priming and the exploitation of stimulus redundancy in perception. We now know that visual objects are perceived more quickly and easily if you have previously been exposed to them, regardless of whether you actually remember having seen them before. This ubiquitous phenomenon, called 'priming', implies that prior exposure to an object changes its representation in the brain (e.g., Bar and Biederman, 1998; Gauthier, 2000; in fact, semantic priming using words is a major experimental paradigm in studies of attention; e.g., Maxfield, 1997). Heller (1988) cites al-Haytham on this (Wang, 1994).

Al-Haytham elaborates on the concept of 'unconscious inference'. He states

> Perception of many of the objects of vision that are perceived by judgment and inference occurs in an extremely short interval of time and in many cases it is not manifest that perception of them occurs by means of judgment and inference because the speed of the inference through which those objects are perceived and the speed of their perception by inference (pp. 130–131).

and

> when it perceives all of them [properties such as shape, size, color, etc.] they become distinct to it at the moment they are perceived

followed by

> Similarly the faculty of judgment does not need an appreciable interval of time to perceive the conclusion (p. 131).

and the strong statement

> Most of the objects of vision that are perceptible by inference are, therefore, perceived extremely quickly. And because of the speed of the perception, it does not become apparent in most cases that they are perceived by inference and discernment; and they are perceived quickly because their premises are manifest and that faculty of judgment has become well accustomed to discerning them (p. 132).

Also, it should be underscored in cognitive science research that the time required to perceive has become an important method for analysing the underlying mechanisms (e.g., the Posner paradigm, Posner and Mitchell, 1967).

As Gregory (1991) writes in an editorial dealing with visual illusions, 'Alhazan describes perception as depending upon knowledge and inference – most illusions being false inferences. He is a neo-Helmholtzian! For he saw perception as very much depending on knowledge and (syllogistic) inference'.

VISUAL ACUITY

Classically, visual acuity was measured in terms of distance (compare with our modern-day testing of Snellen visual acuity) – that is, move an object farther and farther away until it can no longer be seen. In modern terminology, we can say that this is a form of the 'minimum visible' criterion for visual acuity (e.g., Lakshminarayanan, 2011) determination.

Ibn al-Haytham was cognisant of the fact that even near objects which are small cannot be seen and that the ability to see small detail varies among individuals. It should be noted that it was not until the time of Robert Hooke (1674) that experimental studies on visual acuity were performed. These experiments are described by Birch (1757). Ibn al-Haytham also recognized the role of contrast and illumination in the determination of visual acuity. Very early on in Book I, al-Haytham describes acuity as well as the role of illumination, color and inter-subject variation:

> Further, we find that sight does not perceive any visible object unless the object is of a certain size (p. 9).

He further goes on to say

> .... shows that the distances from which a visible object can be perceived and the distances at which it disappears depends on the size of the object (p. 10).

And after a description of a gedanken experiment:

> In terms of luminance, a brighter object can be seen at a given distance where a less brighter object both of the same size disappears

and

> distances from which sight can perceive visible objects and the distances at which they become invisible vary with the lights existing in those objects (p. 10).

and concludes,

> It is thus evident that sight does not perceive any object unless the object has some light in it (p. 12).

> In terms of color and contrast, he writes

> We also find that brilliant white and bright colored bodies are visible from distances at which dull, earthy and dark bodies disappear from view even when the bodies are identical in size and light and all other conditions except color (p. 10).

and he then elaborates that when one approaches from a distance objects on ground of equal (or approximately equal) size and colors (some brilliant white, others bright and dark colors), which are illuminated by the same light,

> bright colored objects will appear before those of earthy or dull colors. Then as he comes nearer still, the others will become visible, until they are all apparent

and concludes

> ...the distances from which objects can be seen and the distances at which they cease to be visible are according to the object's color (p. 10).

> What is the role of resolution (our more conventional way determining acuity)? Here he writes,

> if the eye acquires a true perception of the object at one of these distances, then moves away from it gradually in an orderly manner, those small parts and fine features ..like designs, incisions, creases or dots will disappear before the object disappear as a whole and the smaller and smaller and finer among these features will disappear before those that are larger and more gross. The distances at which the small parts become invisible and the fine features confused and indistinct are found to be many, indeterminate and unlimted (pp. 11–12).

> Al-Haytham attributes variations among individuals to 'strength or weakness of sight' and

> for some small bodies are perceived and sensed by some people but cannot in any way be seen by many others whose sight is not very strong (p. 10).

and in concluding a discussion of why the eye cannot perceive a visible object unless it is greater than a particular size, he writes this sensation is not without limit but [extends only] to the limit that the power of sense may reach and it also varies in [different] eyes according to their different powers, for some eyes are more powerful than others (p. 107).

Before concluding this section on visual acuity, it is worth noting that al-Haytham seems to have anticipated the 'preferential looking' paradigm to investigate vision in infants and small children (Teller et al., 1974; Atkinson et al., 1982). The method as currently used is based on the work of the developmental psychologist Fantz (1965) and uses a two alternative forced choice psychophysical testing procedure. In his Book II, on page 136 al-Haytham discusses perceptual discrimination in the context of unconscious inference. He writes:

clear evidence that man naturally makes inferences without at the same time being aware of making them…is furnished by what can be observed in children in their early development: for a child in early development and at the beginning of awareness perceives many of these things which a man of perfectly developed judgment perceives.

He then goes on to describe the result of a two alternative forced choice procedure:

if a child who is not extremely young nor of perfectly developed judgment is shown two things of the same kind, say two rare fruits or garments or such things as children like and is made to choose between them, assuming one of them is beautiful in appearance and other ugly, then he will choose the beautiful and refuse the ugly one

Again if he is made to choose between two things of the same kind which are both beautiful but one of which is more beautiful than the other, he will often choose the more beautiful object even though the other is also beautiful…. Now the child's preference for the beautiful over the ugly can only be made by comparing one with the other….. .(if he makes such a choice) can only take place after he has compared the two with one another and after perceiving the form of each…But preferring…can only be due to universal premise "what is more beautiful is better and what is better is more worthy of choice".

The last sentence is nothing but the fundamental concept in preferential looking – the child will prefer to look at the window where there is detail (spatial frequency grating) and not an empty grey field (when the grating spatial frequency has gone beyond the visual acuity limit of the infant).

CONCLUSION

The philosopher of science, Thomas Kuhn, in a reference to scientific textbooks, writes that they 'are the unique repository of the finished achievements of modern physical scientists' (Kuhn, 1977). This statement applies to the Kitab al-Manazir and its author. As noted previously, there are seven volumes of the Kitab. However, only the first three volumes have been translated into English, one by Sabra (1989) and the second from the text *De Aspectibus*, the medieval Latin version of the Kitab by Smith (2001). One wonders how many other insights and discoveries are hidden in the other volumes; for example, in Volume 7, al-Haytham deals with the moon illusion (Section 'The Moon Illusion').

In this chapter, I have dealt with some of the many visual phenomena discussed in the first three books of the Kitab. Even though al-Haytham's contributions to optics are rather well known, his contributions found in the other volumes are not extensively cited. The two exceptions are the papers by Bauer (1911) and the more recent one by Howard (1996). Al-Haytham's work on visual perception anticipated and dealt with many topics in physiological optics/vision science. Hence, in Section 'Introduction', I asserted that Ibn al-Haytham should be considered as the founder of physiological optics.

More recently, Fryczkowski et al. (1996) have argued that a Polish scientist, Erazmus Ciołek Witelo (1213–) should be considered as the father of physiological optics. Witelo wrote an important ten volume manuscript, Peri-Optikes, which was first published in 1535 under the title: *'Vitellionis mathematici doctissimi ... vulgo Perspectivam vocant libri X,'* which is translated to English as *Perspectiva*. These authors argue that even though many of Witelo's ideas were similar to those of al-Haytham, Witelo developed his own interpretation of many optical issues. They further state, 'Witelo's achievements and contributions to the understanding of optical and anatomical aspects of the eye and visual function have remained unknown in western literature. Undoubtedly, Witelo should be recognized as the father of physiological optics.'

That being said, there is no question that Witelo's work still rests on the deep foundation laid by al-Haytham.

Some of the visual phenomena discussed in the Kitab were already known, for example, to Ptolemy. However, al-Haytham expanded on these, did careful observations and came to important conclusions. Many of these results were rediscovered by later European (and other) scientists. Without doubt, al-Haytham established the unification of the physical and psychological aspects of vision and integrated the mathematical, physical, physiological and psychophysical domains in order to establish the field of vision science/Physiological optics.

REFERENCES

C Aaen-Stockdale, 2008, Ibn Al Haytham and psychophysics, *Perception*, 37:636–638.

F Aguilonlus, 1613, *Opticorum Libri VI Antwerp* (cited Helmholtz, 1910, vol. III, p. 183).

J Al-Khalili, 2015, In retrospect: Book of optics, *Nature* 518:164–165, doi:10.1038/518164a.

J Atkinson, O Braddick and E Pimm-Smith, 1982, Preferential looking for monocular and binocular testing of infants, *Br. J. Ophthalmol.*, 66:264–268.

T Bahill and L Stark, 1978, The trajectories of saccadic eye movements, *Sci. Am.*, 240:1–12.

M Bar and I Biederman, 1998, Subliminal visual priming, *Psychol Sci.*, 9:464–468.

H Bauer, 1911, Die Psychologie Alhazens, *Beitrage zur Geschichte der Philosophie des Mittelalters, 1911*, 10(5):29–32.

N Bhatia and SSCS Vandana, 2010, Survey of nearest neighbor techniques, *Int. J. Comp. Sci. Info. Sec.*, 8:302–305.

T Birch 1757, *The History of the Royal Society of London*, vol. 3, Millar, London, p. 120, 1757.

E G Boring, 1943, The moon illusion, *Am. J. Phys.*, 11:55–60.

K C Ciuffreda and L Stark, 1975, Descartes's law of reciprocal innervation, *Am. J. Optom. Physiol. Optics*, 52:663–673.

M Clark and L Stark, 1975, Time optimal behavior of human saccadic eye movement, *IEEE Trans. Automatic Control*, 20:345–348.

A J Cogan, 1979, The relationship between the apparent vertical and the vertical horopter, *Vision Res.*, 19:655–665.

T M Cover and P E Hart, 1967, Nearest neighbor pattern classification, *IEEE Trans. Information Theory*, IT-13:21–27.

F Crick, D Marr and T Poggio, 1980, An information processing approach to understanding the visual cortex. Chapter 21 In: *The Cerebral Cortex*, ed. F O Schmitt and F G Worden, MIT Press, Cambridge, MA, pp. 505–531.

K Duncker, 1929, Über induzierte Bewegung (Ein K Beitrag zur Theorie optisch wahrgenommener Bewegung), *Psychologische Forschung*, 12:180–259.

R Fantz, 1965, Visual perception from birth as shown by pattern selectivity, *Ann. NY Acad. Sci.*, 118:793–814.

G T Fechner, 1860, *Elemente der Psychophysik*, 2 volumes, Druck and Verlag Bratkupf und Hartel, Leipzig.

R P Feynman, 1985, Commencement lecture, California Institute of Technology, 1974. *Surely You're Joking Mr. Feynman*, Knopf, New York, p. 343.

S Franz, 1899, After-images, *Psych Reviews: Monograph Suppl.* 3:1–61.

A W Fryczkowski, L Bieganowski and C N Nye, 1996, Witelo--Polish vision scientist of the middle ages: Father of physiological optics, *Surv. Ophthalmol.*, 41:255–260.

I Gauthier, 2000, Visual priming: The ups and downs of familiarity, *Current Biol.*, 10:R753–R756.

K Gegenfurtner, 2016, The interaction between vision and eye movements. *Perception*, 45: 1–25, doi: 10.1177/0301006616657097.

G Gescheider, 1997, *Psychophysics: The Fundamentals*, 3rd ed., Psychology Press, Mahwah, New Jersey.

J J Gibson, 1950, *The Perception of the Visual World*, Houghton Mifflin, Boston.

A Gorea, 2015, A refresher of the original Bloch's law paper (Bloch, July 1885), *i-Perception*, 6, doi: 10.1177/2041669515593043.

R Gorini, 2003, Al-Haytham the man of experience. First steps in the science of vision, *J. Inter Soc. Hist. Islam Med.*, 2(4):53–55.

R L Gregory, 1991, Putting illusions in their place, *Perception*, 20:1–4

A Gullstrand, 1909, The optical system of the eye. Appendix 11.3. In: *Physiological Optics*, ed. Helmholtz H Von, 3rd ed., vol. 1 (Hamburg, Voss, 1909), 350–358.

G Hatfield, 2002, Perception as unconscious inference. In: *Perception and the Physical World: Psychological and Philosophical Issue in Perception*, ed., D Heyer and R Mausfeld. C John Wiley & Sons, Ltd., New York, pp. 115–143.

D Heller, 1988, History of eye movements. In: *Eye Movement Research*, ed., G Luer, U Lass and J Shallo-Hoffman, Hogfre, Toronto, ON., pp. 37–54.

M E Hellman, 1970, The nearest neighbor classification rule with a reject option. *IEEE Trans. Sys. Sci. Cybernetics*, SC-6, 179–185.

H v Helmholtz and J P C Southall, 1910/1962, *Helmholtz's Treatise on Physiological Optics*, Dover Publications, New York.

M Hershenson, 1989, *The Moon Illusion*, Erlbaum, Hilldale, New Jersey.

V Holst and H Mittelstaedt, 1950, Das Reafferenzprinzip, *Naturwissenschaften*, 37:464–476.

I Howard, 1996, Allhazen's neglected discoveries on visual phenomena, *Perception*, 25:1203–1217.

I Howard and N J Wade, 1996, Ptolemy's contribution to the geometry of binocular vision, *Perception*, 25:1189–1206.

S Joseph and V Lakshminarayanan, 2001, Analytic geometric representation of the vertical horopter, http://www.aaopt.org/analytic-geometric-representation-vertical-horopter.

L Kaufman and J H Kaufman, 2000, Explaining the moon illusion, *Proc. Natl. Acad. Sci., USA*, 97:500–505, doi:10.1073/pnas.97.1.500.

L Kaufman and I Rock, 1962, The moon illusion I, *Science*, 136:953–961.

O Khaleefa, 1999, Who is the founder of psychophysics and experimental psychology? *Am. J Islamic Soc. Sci.* 16:1–26.

O Khaleefa and H Manna, 2000, Ibn al-Haytham studies of visual illusions: New discoveries in the history of experimental psychology. International Journal of Psychology, Abstracts of the xxvll International Congress of Psychology. Stockholm, Sweden 23–28 July 2000.

O Khaleefa and H Manaa, 2001, Ibn al-Haytham studies of visual illusions: New discoveries in the history of experimental psychology, *Umm Al-Qura University J.*, 13:19–48 (Saudi Arabia).

F Kingdom, 1997, Simultaneous contrast: The legacies of Hering and Helmholtz, *Perception*, 26:673–677.

E Kirchner, 2015, Color theory and color order in medieval Islam: A review, *Color Sci. Applications*, 40:5–16.

L Koenigsberger, 1965, *Hermann von Helmholtz* (translated by Frances A Welby), Dover, New York.

K Koffka, 1935, *Principles of Gestalt Psychology*, Harcourt, Brace, New York.

T Kuhn, 1970, *The Structure of Scientific Revolutions*, University of Chicago Press, Chicago.

T Kuhn, 1977, The function of measurement in modern physical science. In: *The Essential Tension: Selected Studies in Scientific Tradition and Change*, pp. 178–224, University of Chicago Press, Chicago.

V Lakshminarayanan, 2011, Visual acuity. Chapter 2.1.3, *Handbook of Video Technology*, ed., J Chen et al., Springer-Verlag, Berlin.

V Lakshminarayanan and J M Enoch, 2010, Biological waveguides, In: *Handbook of Optics*, ed., J M Enoch and V Lakshminarayanan, vol. 3, Optical Society of America, Washington, DC.

V Lakshminarayanan, A K Ghatak and K Thyagarajan, 2001, *Lagarangian Optics*, Kluwer, Dordrecht, Netherlands.

A Lejeune, 1948, *Euclide et Ptolemee, deux stades de l'optique geometrique grecque*, Bibliotheque de l'Universite, Louvain, pp. 89–95.

A Lejeune, 1956, *L'optique de Claude Potolemee, dans la version latinee d'apres l'arabe de l'emir Eugene de Sicile*, Lovaine, Repr., Leiden/New York.

H Leopold, J Zelek and V Lakshminarayanan 2017, Deep learning methods applied to retinal image analysis. In: *Biomedical Signal Processing in Big Data*, eds., E Sejdic and T Falk, CRC Press, Boca Raton, FL, 2017

D C Lindberg, 1967, Alhazen's theory of vision and its reception in the West, *Isis*, 58:321–341.

D C Lindberg, 1976, *Theories of Vision: From Al Kindi to Kepler*, The University of Chicago Press, Chicago, IL.

Z-L Lu and B Dosher, 2013, *Visual Psychophysics: From Laboratory to Theory*, MIT Press, Cambridge, Massachusetts.

L Maxfield, 1997, Attention and semantic priming: A review of prime task effects, *Consciou Cognit.*, 6:204–218.

R P McDonald and P T O'Hara, 1964, Size-distance invariance and perceptual constancy, *Am. J Psychol.*, 77:276–280.

S Nanavati, 2009, A history and experimental analysis of the moon illusion, *The New School Psychology Bulletin*, 6:15–25.

National Center for Education Statistics, 2016, https://nces.ed.gov/ipeds/cipcode/cipdetail.aspx?y=55&cipid=87465 (accessed 10/22/2016).

K Ogle, 1965, *Researches in Binocular Vision*, Hafner Publishing Co., New York.

D Park, 1997, *The Fire within the Eye*, Princeton University Press, Princeton, New Jersey.

M Pellilo, 2014, Alhazen and the nearest neighbor rule, *Pattern Recognit. Lett.* 38:34–37.

N Perveen, D Kumar and I Bhardwaj, 2013, An overview on template matching methodologies and its applications, *Int. J. Res. Comp. Comm. Tech.*, 10:988–995.

S L Polyak, 1941, *The Retina*, University of Chicago Press, Chicago, IL.

M I Posner and R F Mitchell, 1967, Chronometric analysis of classification, *Psychol. Rev.*, 74:392–409.

C Ptolemy, 1952, The almagest. (Trans: R C Taliaferro), In: *The Encyclopedia Britannica Great Books of the Western World*, vol. 16, Encyclopedia Brittanica, New York, NY, pp. 1–478.

R Rashed, 2002, A Polymath in the 10th century, *Science*, 297:773, doi:10.1126/science.1074591.

D Reynaud, 2003, Ibn al-Haytham sur la vision binoculaire: un précurseur de l'optique physiologique. *Arabic Sciences and Philosophy*, Cambridge University Press (CUP), Cambridge, UK (13), pp. 79–99.

I Rock, 1983, *Logic of Perception*. MIT Press, Cambridge.

I Rock and L Kaufman, 1962, The moon illusion II, *Science*, 136:1023–1031.

W D Ross, 1931, *The Works of Aristotle*, vol. 3, Clarendon Press, Oxford, UK.

H E Ross and K Bischof, 1981, Wundt's views on sensations of innervation: A reevaluation, *Perception*, 10(3):319–329.

H E Ross and C Plug, 2002, *The Mystery of the Moon Illusion*, Oxford University Press, New York.

G Russell, 1996, The emergence of physiological optics. In: *Encyclopedia of the History of Arabic Science*, eds. R Rashed and R Morelon, vol. 2, pp. 672–715, Routledge, London.

A I Sabra, 1964, Explanation of reflection and refraction: Ibn al Haytham, Descartes and Newton. In: *Actes de dixieme congress international d'histoire des sciences*, ed., H Guerlac, Ithaca, vol. I, pp. 551–554.

A I Sabra, 1966, Ibn al-Haytham's criticisms of Ptolemy's Optics, *J. Hist. Philosophy*, 4:145–149.

A I Sabra, 1972, Ibn al Haytham, Abu 'Ali al_hasan ibn al-Hasan, *Dict. Sci. Biography*, 6:189–210.

A I Sabra, 1978, Sensation and inference in Alhazen's theory of visual perception. In: *Studies in Perception*, eds., P K Machamer and R G Turnbull, Ohio State University Press, Columbus, OH. 217–247.

A I Sabra, 1987, Psychology versus mathematics: Ptolemy and Alhazen on the moon illusion. In: *Mathematics and its Application to Science and Natural Philosophy in the Middle Ages*, ed., E Grant and J Murdoch. Cambridge University Press, Cambridge.

A I Sabra, ed., 1989, *The Optics of Ibn al-Haytham. Books I-II-III: On Direct Vision.* (English Translation and Commentary. 2 vols, Studies of the Warburg Institute, 40, translated by Sabra, A I) The Warburg Institute, University of London, London.

L M Sa'di, 1956, Ibn al-Haytham (Alhazen), Medieval Scientist, *University of Michigan Medical Bulletin*, 22:249–257.

R A Schachar, 1994, Zonular function: A new hypothesis with clinical implications, *Ann. Ophthalmol.* 26:36–38.

J Schmidhuber, 2014, Deep learning in neural networks: An overview, *Tech Report IDSIA 03 14*, ArXiv. 1404.7828v4.

E L Schwartz, 1977, Spatial mapping in the primate sensory projection: Analytic structure and relevance to perception, *Biol. Cybern.* 25:181–194.

S Se and M Brady, 2002, Ground plane estimation, error analysis and applications, *Robotics Autonomous Systems*, 39:59–71.

R Shepard and J Metzler, 1971, Mental rotation of three dimensional objects, *Science*, 171:701–703.

A M Smith, 1986, Ptolemy's Theory of Visual Perception. *Trans. Am. Phil. Soc.*, vol. 86, American Philosophical Society, Philadelphia, Pennsylvania, pp. 109–110.

G Smith, 1995, Schematic eye models: History, description and applications, *Clinical Experimental Optometry*, 78:176–189.

A M Smith, 2001, *Alhacen's Theory of Visual Perception*, American Philosophical Society, Philadelphia, Pennsylvania.

W S Stiles and B H Crawford, 1933, The luminous efficiency of light entering the eye pupil at different points, *Proc. Royal Soc. London, Ser B*, 112:428–450.

A Taha, 1990, Psychophysics according to Ibn al-Haytham (abstract only), *Arab J Psychiatry*, 1:3.

B Tatler, N Wade, H Kwon, J Findlay and B Velichkovsky, 2010, Yarbus, eye movments and vision, *i-Perception*, 1:7–27, dx.doi.org/10.1068/i0382.

D Y Teller, R Morse, R Borton and D Regal, 1974, Visual acuity for vertical and diagonal gratings in human infants, *Vision Res.*, 14:1433–1439.

G J Toomer, 1964, Review: Ibn al-Haythams Weg zur Physik by Matthias Schramm, *Isis*, 55(4):463–465, doi:10.1086/349914.

R B H Tootell, M S Silverman, E Switkes and R L De Valois, 1982, Deoxyglucose analysis of retinotopic organization in primate striate cortex, *Science*, 218:902–904.

A Treisman, 1985, Preattentive processing in vision, *Computer Vision, Graphics, Image Processing*, 31:156–177.

D Tweed, W Cadera and T Vilis, 1990, Computing three-dimensional eye position quaternions and eye velocity from search coil signals, *Vision Res.*, 30(1):97–110.

N Unal and M Elcioglu, 2009, Anatomy of the eye from the view of Ibn al Haytham (965–1039), *Saudi Med. J.*, 30:323–328.

N J Wade, 1996, Descriptions of visual phenomena from Aristotle to Wheatstone, *Perception*, 25:1137–1175.

Q Wang, P Cavangh and M Green, 1994, Familiarity and the pop-out in visual search, *Perception Psychophysics*, 56:495–500.

A M Wong, 2004, Listing's law: Clinical significance and implications for neural control, *Surv. Ophthal.*, 49(6):563–575.

G S Wasserman, 1978, *Color Vision: A Historical Introduction*, Wiley, New York.

A L Yarbus, 1967, *Eye Movements and Vision*, Plenum Press, New York.

7 Ibn al-Haytham's Problem

Pierre Coullet and Roshdi Rashed

CONTENTS

Ibn al-Haytham's Problem in His Book on Optics 109
    Lemma 7.1.. 110
    Lemma 7.2.. 110
Later Solutions of Ibn al-Haytham's Problem 112
    Trigonometric Solution 113
    Complex Solution... 116
    Tychsen Solution 117
    'Ray-Tracing' Solution 119
References... 121

IBN AL-HAYTHAM'S PROBLEM IN HIS BOOK ON OPTICS

In Books IV and V of *Book of Optics*, Ibn al-Haytham [1] studied systematically and exhaustively the reflection of light on various mirrors. In the fifth book, he raises the issue that bears his Latin name: 'The problem of Ibn al-Haytham'. He begins, in the simplest case, with that of the plane mirror. If two points *A* and *B* and a plane mirror *DE* are given, how to determine the point of reflection of a light ray emitted from *A*, reflected to *B*. One must then find a point *C* on the mirror so that the straight lines *AC* and *BC* represent the incident and the reflected rays. Later, Ibn al-Haytham provides a mechanical model of the mechanism of light reflection. Given a circular pool, find all paths from a ball *A* hitting a ball *B* after a single reflection on the pool table. This model is the general case encountered by Ibn al-Haytham immediately after the simple case, that is to say when the mirror is spherical, cylindrical or conical either concave or convex. The successors of Ibn al-Haytham resumed this issue, from the Arabic text by Kamāl al-Din al-Fārisī [2] and Ibn H d ud [3], or from the Latin translation of his *Optics*, published by Risner [4]. Still, this Latin translation is poor and contains some inaccuracies that have been unfairly charged to Ibn al-Haytham. However, Galileo, Huygens, Sluse, Barrow and many others worked from this translation. The situation remained the same until 1942. It was in effect on that date that the physicist and historian of *Optics* M Nazif [7] wrote a study of optics of Ibn al-Haytham starting from the Arabic manuscripts. Recently, A Sabra [13,14] gave the edition of books IV and V of the Treaty. To pose and solve the problem that bears his name, Ibn al-Haytham outlines and demonstrates six lemmas. We will discuss here the first two.

LEMMA 7.1

'Given the known circle *ABC*, in which the diameter *BC* has been produced on the *C* side, given the segment *KI* and the point *A* on the circumference of the circle, we want to draw, from *A*, a line *AHD* such that its part which lies between the diameter and the circle, homologous to the line *HD*, be equal to the segment *KI*' [5] (Figure 7.1).

So there are several cases according that *AB=AC* and whether *AHD* is tangent to the circle or cuts the circle in *A* and *H*. Ibn al-Haytham successively examines these cases.

LEMMA 7.2

'Let a circle *ABC*, the diameter *BC*, a point *A* on the circumference of the circle and given a segment *GH*. We want to draw from point *A* a line that intersects *BC* and ends on the circle, such that its portion between the circle and the diameter is equal to *GH*' [6] (Figure 7.2).

To establish these two lemmas and construct the point *D*, Ibn al-Haytham proceeds by the intersection of a rectangular hyperbola and a circle (Lemma 7.1) (Figure 7.3), and by the intersection of two branches of a hyperbola and a circle (Lemma 7.2). Every time, he mentions the propositions of *Conics* of Apollonius that he is using. As the ideas of the demonstration of the two lemma are the same, we discuss here only the case of Lemma 7.1 [7].

Ibn al-Haytham starts by giving himself a segment *IN* (see Figure 7.4) and draws *BA*, *CA* and the parallel *CG* to *AB* (see Figure 7.3). Let us construct the rectangle *IQNM* such that $\overparen{INL} = \overparen{DCA}$ and $\overparen{INM} = \overparen{DCG}$ (see Figure 7.4). One draws the rectangular hyperbola which passes though *M* with asymptotes *OQ* and *QL*. We cut

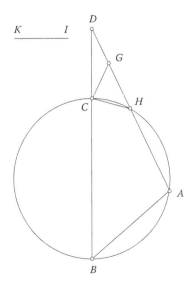

FIGURE 7.1   Lemma 7.1

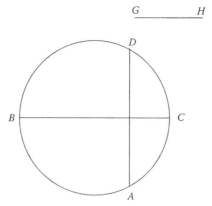

FIGURE 7.2 Lemma 7.2.

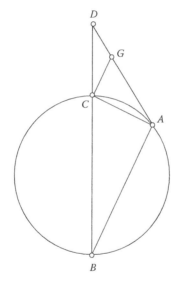

FIGURE 7.3 Lemma 7.2.

this hyperbole by a circle with center M and radius $IN(BC/KE)$; U is the intersection point. We draw MU, which meets the asymptotes with $OM=LU$. The parallelogram $LMIP$ is constructed (see Figure 7.4). One then shows that

$$\frac{AG}{CD} = \frac{BC}{KE}$$

As the two lemmas are similar, much more, are two forms of the same geometric method, Nazif has developed a single demonstration for the two lemmas at a time, using only the ideas of Ibn al-Haytham. Here is this proof:

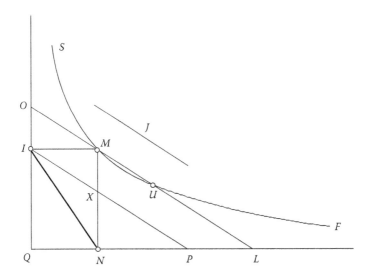

FIGURE 7.4 Lemma 7.2.

Let a point A on the circumference of a circle and a diameter DBC (see Figure 7.5). We want to draw a line from A, which cuts the circle in DBC – or its extension to E – such as $DE=FG$, where FG is a given segment. Let us draw CH parallel to AB which cuts the circle in H. From H, draw a hyperbola whose asymptotes are the extensions of AC and AB. Determine the line HI such that

$$HI = \frac{BC^2}{FG}$$

Draw the circle with centre H and radius HI. This circle cuts the hyperbola in four points I, I_1, I_2 and I_3, and let us join H and I, H and I_1, H and I_2, H and I_3. From A, draw four line parallels to these segments such that each of these lines cuts the circle circumscribed to ABC and the diameter or its extension. One shows that each of these lines solves the problem.

LATER SOLUTIONS OF IBN AL-HAYTHAM'S PROBLEM

The question raised by Ibn al-Haytham is that of the multiple images in a curved mirror. Let $A=(x_A,y_A)$ be the position of a point observer and $B=(X_B,y_B)$ the position of the source also supposedly be a point. Let us assume that the roles of A and B can be inter-changed without inconvenience. Ibn al-Haytham and later Huygens show that the solutions of this problem can be obtained as the intersection of a rectangular hyperbola and a circle mirror. Many mathematicians have tackled Ibn al-Haytham's problem, including Galileo, Sluse, Huygens, Barrow, Quetelet, Tychsen and, most recently, Neumann [8] who showed, in particular, the impossibility of solving it with a compass and ruler. The most comprehensive analysis has been given by C. Tychsen [12]. Remarkably, it solves the problem by explicitly giving its bifurcation set.

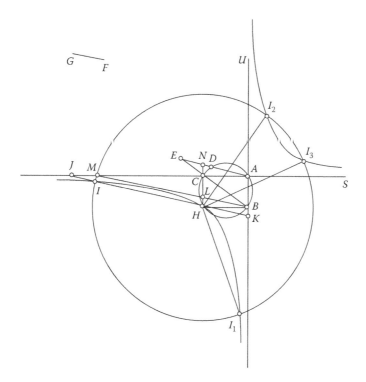

FIGURE 7.5 Lemma 7.2.

TRIGONOMETRIC SOLUTION

We first follow the solution given by Heinrich Dorrie in the classic book *100 Great Problems of Elementary Mathematics* [9]. In Figure 7.6, $\alpha = \widehat{D'A'M}$, $\beta = \widehat{D'B'M}$, $\phi = \widehat{D'B'M}$ and $\theta = \widehat{MOB'} = \widehat{MA'O}$. The law of reflection leads to

$$\theta = \beta - \phi = \phi - \alpha \tag{7.1}$$

Let $M=(x,y)$, $A=(x_A,y_A)$, $B=(x_B,y_B)$. The mirror, without loss of generality, is assumed to be the unit circle. We then have

$$\tan \phi = \frac{y}{x}$$

$$\tan \alpha = \frac{y_A - y}{x_A - x}$$

$$\tan \beta = \frac{y_B - y}{x_B - x}$$

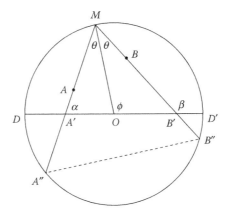

FIGURE 7.6   Geometry of the problem of Ibn al-Haytham.

The condition of reflection (7.1)

$$\tan(\beta - \phi) = \tan(\phi - \alpha)$$

and the trigonometric identity

$$\tan(a+b) = \frac{\tan(a)\tan(b)}{1 + \tan(a)\tan(b)}$$

lead to the following relation between x and y:

$$\frac{xy_A - yx_A}{x^2 + y^2 - xx_A - yy_A} = -\frac{xy_B - yx_B}{x^2 + y^2 - xx_B - yy_B}$$

or, after simplifications,

$$p(x^2 - y^2) - 2qxy + (x^2 + y^2)(ry - sx) = 0 \qquad (7.2)$$

where

$$p = x_A y_B + x_B y_A$$

$$q = x_A x_B - y_A y_B$$

$$r = x_A + x_B$$

$$s = y_A + y_B$$

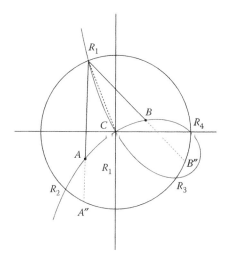

FIGURE 7.7   Barrow's solution.

Equation 7.2 defines a cubic curve whose intersections with the circle are the solutions of the problem of Ibn al-Haytham (see Figure 7.7). This is the solution proposed by Barrow (see Reference 15). The polar representation of the Barrow's curve write

$$\rho = c\,\frac{\cos(2(t+\delta))}{\cos(t+\eta)}$$

where

$$c = \sqrt{\frac{p^2+q^2}{r^2+s^2}}$$

$$\tan(\delta) = \frac{p}{q}$$

$$\tan(\eta) = \frac{r}{s}$$

Let us note that in Equation 7.2, x^2+y^2 can be replaced with 1, since we are looking at the intersections of Barrow's curve with the unit circle. This equation then describes a quadratic curve, namely that of an equilateral hyperbola. This is the hyperbola of Ibn al-Haytham and Huygens. Its intersections with the circle are the solutions of the problem of Ibn al-Haytham (Figure 7.8).

$$p(x^2-y^2)-2qxy+ry-sx=0 \tag{7.3}$$

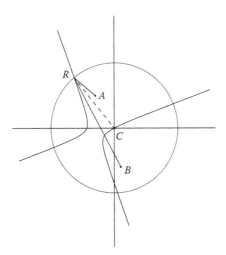

FIGURE 7.8   Hyperbola of Ibn al-Haytham and Huygens.

Complex Solution

The above analysis can be greatly simplified by using the complex plane [15].

Let $a = x_A + iy_A$, $b = x_B + iy_B$ and $z = x + iy$ respectively the complex coordinates of A, B and P. The equality of incident and reflected angles reads

$$\theta = \arg\left(\frac{a - z}{z}\right) = -\arg\left(\frac{b - z}{z}\right)$$

or

$$\arg\left(\frac{(a - z)(b - z)}{z^2}\right) = 0$$

and equivalently

$$\mathrm{Im}\left(\frac{(a - z)(b - z)}{z^2}\right)$$

Using the equation of the unit circle $z\bar{z} = 1$, we finally get the complex version of Equation 7.3

$$\mathrm{Im}((ab)\bar{z}^2) = \mathrm{Im}((a + b)\bar{z})$$

where

$$p = \mathrm{Im}(ab)$$

$$q = \mathrm{Re}(ab)$$

$$r = \mathrm{Re}(a+b)$$

$$s = \mathrm{Im}(a+b)$$

An appropriate choice of coordinates simplifies the expression of hyperbola. Precisely, if one chooses the axis of x as the bisector of the angle ACB, we get $p = \mathrm{Im}(ab) = 0$; then the equation of the hyperbola becomes

$$xy - h_y x - h_x y = 0 \tag{7.4}$$

where $H = (h_x, h_y)$ is the centre of the hyperbola

$$H = \left(\frac{r}{2q}, -\frac{s}{2q} \right)$$

Let us note that the problem of Ibn al-Haytham is reduced to a two-parameter problem, h_x and h_y, which are the coordinates of the centre of the equilateral hyperbola, while it seems to be a four-parameter problem $(x_A, y_A, x_B, y_A, y_B)$. The elimination of y from Equation 7.4 and the equation of the circle $x^2 + y^2 = 1$ leads to the quartic equation

$$x^4 - 2h_x x^3 + (h_x^2 + h_y^2 - 1)x^2 + 2h_x x - h_x^2 = 0$$

At this stage of the analysis, one could consider that the problem is solved analytically, since the solutions of quartic equations are given by an explicit formula. These expressions are, however, useless to study the bifurcations as h_x and h_y vary.

TYCHSEN SOLUTION

The elegant solution proposed by Tychsen [12], a Danish mathematician in 1861, allows one to push the analysis much further. This solution is mentioned by Henning Bach in a report found on the Internet entitled 'Some Ray Tracing Problems Related to Circles and Ellipses' [10]. Let us introduce the following parameterisation of the circle

$$x = \frac{1-t^2}{1+t^2}$$

$$y = \frac{2t}{1+t^2}$$

Using these expressions in Equation 7.4, we get another quartic equation

$$t^4 + at^3 + bt - 1 = 0 \tag{7.5}$$

where

$$a = -2\left(\frac{1+h_x}{h_y}\right)$$

$$b = 2\left(\frac{1-h_x}{h_y}\right)$$

Equation 7.5 has two or four real solutions [12]. The bifurcation set where two solutions coincide occurs when

$$\left(\frac{a-b}{4}\right)^{2/3} - \left(\frac{a+b}{4}\right)^{2/3} = 1 \tag{7.6}$$

The locus of points in the plane $(a-b)$ where the problem of Ibn al-Haytham has double solutions is achieved by simultaneously seeking zeros of the quartic equation and its derivative

$$4t^3 + 3at^2 + b = 0 \tag{7.7}$$

From Equations 7.5 and 7.7, one can compute a and b as a function of t

$$a = \frac{\left(t^2 - 1\right)^3}{8t^3}$$

$$b = -\frac{\left(t^2 + 1\right)^3}{8t^3}$$

The elimination of t between these two expressions leads to condition (7.6). In the variables of the problem of Ibn al-Haytham, this condition simply becomes

$$h_x^{(2/3)} + h_y^{(2/3)} = 1 \tag{7.8}$$

In the plane $(x-y)$, the curve (see Figure 7.9)

$$x^{(2/3)} + y^{(2/3)} = 1$$

defines the bifurcation set. When the centre of the hyperbola is located inside curve, the problem of Ibn al-Haytham has four real solutions and only two when the centre

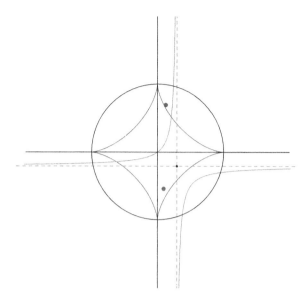

FIGURE 7.9   Tychsen's astroid.

is outside. The regular part of the corresponds to a tangential bifurcation where two solutions appear or disappear. The four cusps correspond to the case where three solutions coincide.

'Ray-Tracing' Solution

We have presented two ways to reduce the problem of Ibn al-Haytham to a quartic equation. They are closely tied to the properties of the circle. Let us consider another way to solve this problem which has the advantage of applying itself in cases where the mirror is no longer circular. Let C be a parametrically defined curve

$$x = f(t)$$

$$y = g(t)$$

Let $\hat{n}$ be the unit normal at a point $M = (f(t), g(t))$

$$\vec{n} = \left(-\frac{g'(t)}{\sqrt{f'^2 + g'^2}}, \frac{f'(t)}{\sqrt{f'^2 + g'^2}} \right)$$

where f' and g' are respectively the derivatives of f and g with respect to t. Let $A = (x_A, y_A)$ the position of the source (see Figure 7.10). Let $\vec{r} = \overrightarrow{MA'}$ be the normalised reflected ray.

$$\overrightarrow{MA} + \overrightarrow{MA'} = 2\overrightarrow{MH} = 2(\overrightarrow{MA}.\hat{n})\hat{n}$$

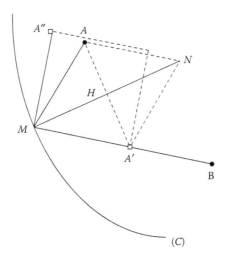

FIGURE 7.10 Ray-tracing.

or

$$\vec{r} = 2(\overrightarrow{MA}.\hat{n})\hat{n} - \overrightarrow{MA}$$

Let us introduce the vector $\vec{s} = \overrightarrow{MO}$ orthogonal to $\vec{r}$

$$\vec{s} = (-r_y, r_x)$$

where $\vec{r} = (r_x, r_y)$. Let $B = (x_B, y_B)$ the position of the observer. The condition for the reflected ray corresponding to the incident ray AM to pass through B is given by

$$A(t; x_A, y_A; x_B, y_B) = \overrightarrow{MB}.\vec{s} = 0 \qquad (7.9)$$

The solutions of this equation, for A and B given, solve the Ibn al-Haytham problem for the mirror C. Let us remark that when $A = 0$ and $A' = 0$, where A' is the derivative of A with respect to t, the problem of Ibn al-Haytham admits a double solution. Similarly, if simultaneously $A'' = 0$, the problem has a triple solution.

Let us look now at how this method inspired by the techniques of 'ray-tracing' [16] allows one to solve the problem of Ibn al-Haytham. The curve $C = (x, y)$ then is the unit circle, $x^2 + y^2 = 1$. Its normal is

$$\hat{n} = (-x, -y)$$

The reflected ray $\vec{r}$ is given by

$$\vec{r} = 2 p\hat{n} - (x_A - x, y_A - y)$$

where

$$p = 1 - x_A x - y_A y$$

This expression has been simplified, using $x^2 + y^2 = 1$. The Ibn al-Haytham condition (7.9) then reduces itself, in the frame such as $x_A y_A = -x_B y_B$, that is when the x bisects the angle $\widehat{AOB}$, to the rectangular hyperbola given by Equation 7.4.

REFERENCES

1. Ibn al-Haytham, *Kitāb al-Manāzir (The Book ? of Optics)*, Books IV–V, The Arabic Text, ed. Sabra A, Kuwait, 2002, p. 194.
2. Kamāl al-Dīn al-Fārisī, *Tanqih al-Manāzir*, ed. al-Maarif Dairat, Hyderabad, India, A929, vol. 1, pp. 458–467.
3. Hogendijk J, Al-Mū'taman ibn Hūd, 11th century King of Saragossa and brilliant mathematician, *Historia Mathematica*, 22, 1–18, 11–12, 1995.
4. *Opticae Thesaurus Alhazeni Arabis Libri Septem*, Johnson Reprint Corporation, 1972 (reprint de l'édition de Federico Risner, Basle, 1572), pp. 143–145.
5. Ibn al-Haytham, op. cit., p. 235.
6. Ibid., p. 240.
7. Nazif M, *Al- Ḥasan ibn al-Haytham, Buḥūlhuku wa-Kūshufuhu'l baṣariyya*, vol. 1, 2, 1942, Fuad I University, Cairo, pp. 496–505.
8. Neumann P M, Reflections on reflection in a spherical mirror, *The American Mathematical Monthly*, 105 (6), 523–528, 1998.
9. Dorrie H, *100 Great Problems of Elementary Mathematics*, Dover, 1965, pp. 197–200.
10. Bach H, *Some Ray Tracing Problems Related to Circles and Ellipses*, Defense Technical Information Center, 1989. http://www.dtic.mil/dtic/tr/fulltext/u2/a219306.pdf.
11. Tychsen C, On a mathematical investigation of a billiard problem, *Matematisk Tidsskrift*, 1859, 60–64.
12. Tychsen C, A remark concerning an equation of the fourth degree, *Matematisk Tidsskrift*, 1851, 141–144.
13. Sabra A I, *The Optics of Ibn al-Haytham. Edition of the Arabic Text of Books IV and V: On Reflection and Images Seen by Reflection*, The National Council for Culture, Arts and Letters, Kuwait, 2002.
14. Sabra A I, Ibn al-Haytham's lemmas for solving Alhazen's problem, *Archive for History of Exact Sciences*, 26, 1982.
15. Smith J D, The remarkable Ibn al-Haytham, *The Mathematical Gazette, The Use of the History of Mathematics in the Teaching of Mathematics*, 76, 1992.
16. Pharr M and Jakob W, *Physically Based Rendering: From Theory to Implementation*, Morgan Kaufmann Publishers, Elsevier, 2010.

8 Ibn al-Haytham and His Influence on Post-Mediaeval Western Culture

Charles M Falco

CONTENTS

Introduction 123
Theories of Vision 124
Appropriation of Ibn-al-Haytham's Work into European Culture 125
Optics 125
Literature 125
Art 126
Religion 128
Conclusion 129
Acknowledgements 129
References 129

INTRODUCTION

Ibn al-Haytham (Latinised as Alhazen or Alhacen) wrote nearly 100 works on topics as diverse as astronomy, poetry and politics. Although today he is largely known for his writings on geometrical optics, astronomy and mathematics, with his landmark seven-volume *Kitāb al-Manāzir* (Book of Optics), published sometime between 1028 and 1038, he made intellectual contributions that subsequently were incorporated throughout Western culture. The *Kitāb al-Manāzir* was translated into Latin as *de Aspectibus* some time prior to 1230 where it profoundly influenced European intellectuals as diverse as the writer Geoffrey Chaucer, the theologian John Wyclif, and the scientific work on optics of Bacon, Pecham and Witelo. The noted science historian David Lindberg wrote that 'Alhazen was undoubtedly the most significant figure in the history of optics between antiquity and the seventeenth century.' However, accurate as that characterisation is, it significantly understates the impact that al-Haytham had on areas as wide-ranging as the art, science, theology and literature of Europe.

THEORIES OF VISION

Translation of ancient writings on the science of optics during the first two centuries of the Islamic Golden Age provided contemporary intellectuals with the theories of vision of Euclid, Ptolemy, Galen and Aristotle. These translations of Greek texts contained disparate views on the way the organ of the eye reacts with light to aid in vision. Although Greek theories were largely appropriated by Islamic scholars until the first quarter of the eleventh century, at that time a new method of inquiry was introduced by Ibn al-Haytham who combined experimental evidence with theoretical insights to develop a remarkably accurate theory of vision. His design and interpretation of experiments became the root source for Western understandings of optics up to the seventeenth century (Lindberg, 1967).

Prior to al-Haytham's work, theories of vision could be broadly classified into one of three categories: extramission, intromission or a combination of the two. Extramission theories, well-known proponents of which are Euclid and Ptolemy, require that some sort of illuminating particles be emitted by the eye (Lindberg, 1981). Although there are obvious flaws with these theories, they do get the geometry correct and there is a one-to-one correspondence between points on the object and points on the eye. Intromission theories by philosophers including Aristotle postulated that objects continuously emitted microscopic replicas of themselves that travelled to the eye of the observer. Although intromission theories avoid some problems of extramission theories, such as near and far objects simultaneously being visible when the eye is opened, they too have obvious flaws. A third class of theories, proposed by Plato and Galen (and Aristotle, to a lesser degree), combined extromission and intromission theories and proposed that light emitted by the eye 'activates' in some way the intervening air to allow replicas, or 'species' to create the image (Lindberg, 1967, 1981).

Ibn al-Haytham, recognising problems with existing theories of vision, proposed a type of intromission theory of vision and validated his proposition by empirical understanding deduced from scientific experimentation. His methodology expanded understanding beyond the theoretical and resulted in him incorporating psychology as well as the behaviour of light and the physiology of the eye to explain vision.

Ibn al-Haytham's experiments included using flat and curved mirrors to control and manipulate light as well as involved observing the effect of light passing through apertures of various sizes into darkened spaces (i.e., *camera obscura*) (Smith, 2001a,b). Significantly, his experiments provided al-Haytham with the theoretical construct of light rays that he used as a means for describing and interpreting the visual system, and linked sight and vision with the properties of light throughout his studies.

One important aspect of vision that al-Haytham did get wrong is the fact that an image projected by a lens is upside down and has its right-to-left parity reversed. Although his experiments with eyeballs should have given him the correct answer, apparently this was more than he could accept in a theory of vision even though it is contained within his optical formalism. Leonardo da Vinci also failed to accept this as an explanation for vision when he approached the problem much later (Kemp, 1977), and ultimately it would be Kepler 500 years later who would follow Ibn al-Haytham's formalism to its logical conclusion when he developed the theory of the retinal image. Significantly, Kepler titled his work 'Amendments to Witelo', incorrectly crediting the thirteenth century priest who had plagiarised al-Haytham's work.

APPROPRIATION OF IBN-AL-HAYTHAM'S WORK INTO EUROPEAN CULTURE

There is little doubt that Ibn al-Haytham's *Kitab al-Manazir*, translated into Latin as *De Aspectibus* in the early to mid-thirteenth century, is largely responsible for the widespread appropriation of its contents by Western European intellectuals (Lindberg, 1967). Although it is important to note that there are variations in organisational structure and interpretation of specific terms between the Latin version and the Arabic original (Smith, 2001a,b), the influence of al-Haytham's writing on European culture was profound, as will be described in the rest of this chapter.

OPTICS

The earliest known reference to the Latin translation of the *Kitāb al-Manāzir*, *De Aspectibus*, is in *De proprietatibus rerum* (On the Property of Things) c1220–1230 by Bartolomeo Anglicus (Smith, 2001a), an early encyclopaedia covering a wide variety of subjects. The text refers several times to the study of optics and to Ibn al-Haytham. This encyclopaedia was required reading at the University of Paris in 1296, available in the university library of the Sorbonne by 1306, and widely used as reference material at Oxford, Cambridge, Canterbury and Merton College by the mid-fourteenth century (Holbrook, 1998).

However, specific proposals contained within *De Aspectibus* are most significantly referred to in the optics manuscripts *Perspectiva* by Roger Bacon (c1268), *Perspectiva* by Erazmus Witelo (c1278) and *Perspectiva communis* by John Pecham (c1280) (Lindberg, 1967). Although today we think of these scholars as optical scientists, they were theologians who approached their work as such, which influenced their interpretation of mediaeval optical theories. Bacon, for example, was a Franciscan friar who secretly transmitted his scientific manuscripts to the Papal court (Smith, 2005). Pecham and Witelo also were priests and relied on Ibn al-Haytham in constructing their own evolving optical theories, although taking liberties with their interpretations and infusing their writings with spiritual undertones. The collective work of these scholars had enormous influence on the progression of optical understanding in Europe in the centuries that followed.

LITERATURE

European literature published during the period after al-Haytham's work had been translated into Latin illustrates how widespread interest and understanding of optics had become, as well as how influential this Arab scholar was. Ibn al-Haytham is referred to several times in the epic poem *Roman de la Rose* (Romance of the Rose) by Guillaume de Lorris and Jean de Meun, one of the most widely read works in the French language for 300 years after its publication in c1275 (Ilardi, 2007). In the text, the authors describe the properties of mirrors, with the text exhibiting a non-trivial understanding of optics. One short passage from these four pages makes its debt to Ibn al-Haytham (Alhacen) quite clear:

> There [in Alhacen's Observations] he will be able to discover the causes and the strengths of the mirrors that have such marvelous powers that all things that are very small – then letters, very narrow writing, and tiny grains of sand – are seen as so great and large and are put so close to the observers – for everyone can distinguish among them – that one can read them and count them from so far off that anyone who seen it phenomenon and wanted to tell about it could be believed by a man who had not seen it or did not know its causes. This would not be a case of belief, since he would have the knowledge of the phenomenon (Ilardi, 2007, p. 44).

A century later, Geoffrey Chaucer refers to Ibn al-Haytham in *Canterbury Tales*, the first major piece of literature in the vernacular English language. Chaucer, too, was influenced by his understanding of the content of Ibn al-Haytham's works on vision and optics as is clear from the following passage:

> They spoke of Alhazen and Vitello and Aristotle, who wrote of curious mirrors and of perspective glasses, as they know who have heard their books (NeCastro, 2007, p. 3).

Interestingly, although Ibn al-Haytham's contributions were disseminated in Europe through the texts of Bacon, Pecham and Witelo where they influenced a period of scientific discovery in optics, they also informed advancement of visual literacy. The science of optics in turn provided the syntax upon which new spatial understandings were constructed. As Greenstein (1997) states, 'Because vision is a cognitive process involving inner sense and intellect, optics links sight with semantics, semiotics, and theories of the soul. It makes use of such fundamental Aristotelian concepts as form, substance, accident, quality, individual, universal, species, and whatness' (p. 682). As mediaeval optical theories increasingly informed scientific inquiry, interest in how the visual system might be applied to artificial representations of space and spatial perception (i.e., painting and drawing) also became important.

ART

There is yet to be a consensus as to what influence perspectivist theories of vision had on the visual arts up to and throughout the Renaissance. Writing on the influence of optical scientists on visual artists, Klein (1961) states, 'We may observe the widespread conviction that there was a close connection between their disciplines, really an identity' (p. 212). He further states, 'If one can believe Rafaello Maffei, the ancient science of Alhazen and Vitellio now included artistic applications and was almost identified with the fine arts' (Klein, 1961, p. 212). Although Kemp (1990) is less certain that perspectivist theories of vision directly affected the development of linear perspective, a more favourable view can be found in the introduction of Smith's (2001a) English translation of *De Aspectibus*, where he states:

> The representation of visual space in Renaissance art was the expression of a world-view implicit in the Perspectivist analysis of sight, a world view based upon the 'geometrization' of visual space. If, however, Alhacen and his Perspectivist followers taught Renaissance artists to 'see' the world in such spatial terms, those artists in turn taught early modern thinkers to see the world in those same terms and thus to conceive of it as a Euclidean continuum.

Written evidence of this 'Euclidean continuum' is interpretable in the writings of such Renaissance masters as Alberti, Ghiberti and da Vinci, and equally important visual evidence is evident in the actual images created throughout the period. Alberti's most notable work, *On Painting*, c1435, for example, employs a model for vision taken directly from Ibn al-Haytham. Greenstein (1997) validates the parallel between al-Haytham's and Alberti's models of vision by stating, 'Alberti's viewer first sees under aspects (*aspecimus*); then recognises by intuition (*intuentes...dignoscimus*); and finally discerns with greater discrimination (*aspicientes distinctius...discernimus*)' (p. 682). These stages of visual succession imply that Alberti understood spatial perception as a layered and complex cognitive process, one that must account for the perceptual tendencies the viewer applies as a means for interpreting any given scene. There is also support for the idea that the 'Euclidean continuum' informed Alberti's understanding of the visual pyramid (Lindberg, 1981).

Kemp (1990), however, is conservative in his estimate of the influence optics had on transitions in visual art practice evident in painting, stating, 'Medieval optical science created far more problems than it solved for Renaissance artists' (p. 345). For example, he notes that Alberti is explicit about having composed *On Painting* for artists and appears indifferent to which direction visual rays might reach the eye of the viewer, that is, whether it was by intromission or extromission. This is interpreted by Kemp (1977) as a break by Alberti with scientific tradition. Kemp's conclusions, however, do not wholly consider the practical implications which occur when the visual system is oriented and applied to visual art practice, and predate the discoveries of the use of optics by early Renaissance painters by a decade (Hockney and Falco, 2000).

Alberti was sufficiently aware of the debate arising from the perspectivist optical tradition about visual rays to write that such considerations are 'useless' for artistic purposes (Kemp, 1977). Subsequently, Alberti focused his attention instead on concepts relative to spatial disposition and composition, and how these two principles are translated and reoriented as objects on a two-dimensional picture plane. Briefly, Alberti was the first to interrupt the visual pyramid by placing the canvas perpendicular to the visual rays at the vertex of the pyramid. Whether Brunelleschi's panel experiments at the Piazza del Duomo or the Piazza della Signoria in Florence were actually the first experimental illustration of this effect will not be discussed here. However, the latter certainly informed the former (Arnheim, 1978; Kemp, 1990). The necessary level of understanding for providing audiences with instruction of these advanced visual considerations subsequently required a new formalism and language, namely perspective.

Lorenzo Ghiberti attempted a theoretical understanding of the arts a decade after *On Painting* was published, relying heavily on the optical theories of Pecham, Witelo and others, which in turn directly relied on Ibn al-Haytham. That Ghiberti had access to a fourteenth century Italian translation of Ibn al-Haytham is certain, given that entire portions of it are incorporated in Book 3 of his *Commentarii* (Fragenberg, 1986; Greenstein, 1997, etc.). Ghiberti's book was incomplete at the time of his death but is described by Lindberg (1981) as 'The most transparent case of the influence of medieval visual theory on a quattrocentro artist… (Ghiberti) presents a complete survey of the mathematical tradition in optics consisting mainly of excerpts and paraphrases drawn from the perspectivists' (p. 152). Subsequently, by the time da Vinci would consider the works of these same optical scientists, he too would be forced to reconcile for himself the relationship between vision, perception and pictorial representation.

Kemp's (1977) research on da Vinci makes it clear that the artist established an 'increasingly sharp separation between perspective as the science of vision and perspective as a geometrical means for constructing a rational picture space', which da Vinci refers to respectively as 'perspective made by nature' and 'perspective made by art'; or *prospettiva naturale* and *prospettiva accidentale* (p. 147). It is also clear that da Vinci saw shortcomings in Alberti's and Ghiberti's affirming or denying the function of the visual pyramid, because he embarked on an intense period of activity applying a methodology and conducting experiments remarkably similar to those of Ibn al-Haytham (Smith, 2001a).

It seems unusual that so little is known how an artist like Alberti arrived at the principles of linear perspective when he was so familiar with the perspectivist tradition. Nevertheless, a clear language of visual literacy in Europe has been established that begins with the work of Ibn al-Haytham and culminates with Alberti's visual pyramid for artistic production during the Renaissance (Falco, 2005; Hockney, 2001, 2006; Hockney and Falco, 2004; Hockney and Falco, 2005a).

RELIGION

Although today we think of Bacon, Witelo and Pecham as optical scientists, they were Roman Catholic priests who approached their work as theologians. In each case, their interest in optics was motivated by their interest in spiritual vision, which was informed by their religious beliefs.

The onset of the Protestant Reformation is typically dated to 1517 when the priest Martin Luther published his '95 Theses' criticising the Roman Catholic Church. Luther, however, built directly on the theology of the fourteenth century priest John Wyclif who is credited with being the intellectual progenitor of the Reformation ('The Morningstar of the Reformation'). One of his revolutionary activities was to translate the Bible from Latin into English, thereby bypassing the priesthood to make its content directly accessible to believers to interpret for themselves. Even more directly than Bacon, Witelo and Pecham, Wyclif used optics in developing his own interpretation of Christian theology. For example, he classified spiritual vision as 'direct', 'refracted', and 'reflected' in his c1360 *De Logica*. He referred to al-Haytham by name in his 1376 *De Civili Dominio* where he discussed the seven deadly sins in terms of the distortions in the seven types of mirrors that had been analysed in al-Haytham's *De Aspectibus*. Wyclif even used the Arabic word for parabolic mirror, *mukephi*, in his Latin text for his 1382 *De Eucharista*.

Perhaps most significantly, in his 1372 *Sermones II*, Wyclif offered a heretical optical alternative to the Transubstantiation of Christ, the most sacred of the seven sacraments of the Roman Catholic Church. The alternative he proposed, directly based on the optics of the Islamic scholar Ibn al-Haytham, was that Christ's presence in Heaven was so powerful that *mukephi* could simultaneously project Him to multiple locations for Mass. Wyclif's 'optical heresy' was so profound that two separate Popes ordered his remains to be unearthed, burnt at the stake and dispersed in a nearby river. It is not overstating it to say Ibn al-Haytham's work on optics had a non-negligible influence on the development of the Protestant Reformation, and hence on a very significant aspect of European civilisation.

CONCLUSION

As the examples from art, science, literature and even Christian theology in this chapter demonstrate, Ibn al-Haytham's intellectual contributions are intimately threaded throughout the core of post-Mediaeval Western culture. It is indeed unfortunate that individual academic disciplines today are largely unaware of the overall scope of his influence but, with effort, that can change.

ACKNOWLEDGEMENTS

I gratefully acknowledge David Hockney for his invaluable insights on painting through his investigation of over 1000 years of European art.

REFERENCES

Arnheim, Rudolf, 1978. Brunelleschi's peepshow, *Zeitschrift fur Kunstgeschichte*, 41 (1), 57–60 (Deutscher Kunstverlag GmbH Munchen Berlin).

Falco, Charles M, 2005. A Amira, A Bouridane and F Kurugollu (Ed.) Analysis of qualitative images, *Proceedings of the Irish Machine Vision and Image Processing Conference* (Queen's University Belfast), p. 11.

Fragenberg, Thomas, 1986. The image and the moving eye: Jean Pelerin (Viator) to Guidobaldo de Monte, *Journal of the Warburg and Courtald Institutes*, 49, 150–171.

Greenstein, Jack M, 1997. On Alberti's 'sign': Vision and composition in quattrocento painting, *The Art Bulletin*, 79 (4), 669–698. Quote found on p. 682.

Hockney, David, 2001, 2006. *Secret Knowledge: Rediscovering the Lost Techniques of the Old Masters*. Thames and Hudson 2001, Penguin Group 2006.

Hockney, David and Falco, Charles M, 2000. Optical Insights into Renaissance Art, *Optics & Photonics News*, 11, 52.

Hockney, David and Falco, Charles M, 2004. The Art of the science of Renaissance painting, *Proceedings of the Symposium on 'Effective Presentation and Interpretation in Museum'*. National Gallery of Ireland, pp. 1–4.

Hockney, David and Falco, Charles M, 2005a. *Proceedings of Photonics Asia. Optical instruments and imaging: The use of optics by 15th century master painters. SPIE*, 5638, 1.

Hockney, David and Falco, Charles M, 2005b. *Proceedings of IS&T-SPIE Electronic Imaging: Quantitative analysis of qualitative images*, SPIE, 5666, 326.

Holbrook, Sue Ellen, 1998. A Medieval scientific encyclopedia 'Renewed by good printing': Wynkyn de Word's English 'De Proprietatibus Rerum', *Early Science and Medicine*, 3 (2), 119–156 (BRILL).

Ilardi, Vincent, 2007. *Renaissance Vision from Spectacles to Telescopes.* American Philosophical Society. (Quote taken from a translation by Charles Dahlberg (Princeton, 1971) of *Roman de la Rose*), p. 43.

Kemp, Martin, 1977. Leonardo and the visual pyramid, *Journal of the Warburg and Courtald Institute*, 40, 128–149. (The Warburg Institute. Regarding da Vinci's separation of the terms 'perspective made by nature' verses 'perspective made by art', Kemp illustrates that da Vinci considered greatly the differences between the two as they related to optics and art (p. 147 (70)).)

Kemp, Martin, 1990. *The Science of Art: Optical Themes in Western Art from Brunelleschi to Seurat.* Yale University Press.

Klein, Robert, 1961. Pomponius Gauricus on perspective, *The Art Bulletin*, 43 (3), 211–230. (College Art Association.)

Lindberg, David C, 1967. *Alhazen's theory of vision and its reception in the West, ISIS*, 58 (3), 321–341.

Lindberg, David C, 1971. *Lines of Influence in the Thirteenth-Century Optics: Bacon, Witelo, and Pecham, Speculum*, 46 (1), 66–83.

Lindberg, David C, 1981. *Theories of Vision from Al-Kindi to Keppler.* University of Chicago Press, 237. (Regarding the quote by Alberti on the 'useless' aspects of the debate over ray theory for the purposes of painting.)

NeCastro, Gerard(Trans. and Ed.), 2007. *The Squire's Tale.* Online resource: http://www.umm.maine.edu/faculty/necastro/chaucer

Smith, Mark A, 2001a. *Alhacen's Theory of Visual Perception: A Critical Edition*, with English Translation and Commentary, of the First Three Books of Alhacen's 'De Aspectibus', the Medieval Latin Version of Ibn al-Haytham's Kitāb al-Manāẓir: Vol. One, *Transactions of the American Philosophical Society, New Series*, 91 (4). American Philosophical Society.

Smith, Mark A, 2001b. *Alhacen's Theory of Visual Perception: A Critical Edition*, with English Translation and Commentary, of the First Three Books of Alhacen's 'De Aspectibus', the Medieval Latin Version of Ibn al-Haytham's 'Kitāb al-Manāẓir': Vol. Two, *Transactions of the American Philosophical Society, New Series*, (91)5, 339–819. American Philosophical Society.

Smith, Mark A, 2005. *The Alhacian Account of Spatial Perception and its Empistemological Implications, Arab Sciences and Philosophy*, 15, 219–240. Cambridge University Press, Cambridge, UK.

Section II

Light-Based Technologies for the Future

This second section of the book reports selected contributions dedicated to the importance of light-based technologies for our societies.

Azzedine Boudrioua gives a highlight of the main concepts and some recent developments of photonics technology as well as future challenges. Today, light-based technologies include strategic fields like space and military ones and also fields of everyday life such as data storage, medicine and industry. The interest of using the photon rather than the electron comes from the very high optical frequencies of the optical signal which allow a very broad bandwidth and offer an unequalled data transmission capacity. Ultimately, the twentieth century was the century of electronics and the twenty-first century is expected to be the century of photonics. It is interesting to emphasise that this important development of light science and technology is mainly based on the principles of reflection and refraction and their related phenomena. These principles have been the focus of the founding works of Arab scholars during the golden age of Islamic civilisation – in particular, the works of Ibn Sahl and Al Hasan Ibn al-Haytham.

In the past few years, NASA's Curiosity rover has advanced our understanding of the red planet to an unprecedented level. The success of this mission, also known as the Mars Science Laboratory mission, is the result of the work of many great scientists and engineers. In his contribution, Noureddine Melikechi emphasises how, more than 1,000 years ago, Ibn al-Haytham contributed significant ideas that are part of the foundation that led to scientific discoveries and technological advances that support the Curiosity mission. It is now more and more recognised that Ibn al-Haytham contributed significantly to the birth of the scientific method. Ibn al-Haytham's desire to better understand the world around him, as well as his determination to challenge orthodoxies in place at his time, served him and humanity very well. He was the first to suggest that light travels to the eye in rays from different points of an object and can arguably be considered one of the founders of the field of optics.

Light has fascinated humans since time immemorial. While we see huge advances today in relation to light and light-based technologies, one needs to remember that our advancements and understanding of light and optics are built upon accumulated contributions from all corners of the earth. In highlighting the importance of light in our lives today, it is useful to recall how some of the early contributions in understanding the role of light in our lives, for example, how we see, have come to pave the way towards later developments, inventions, methodologies and understanding in science and developments in technology that we still use today. Mohamed El Gomati presents the art of magnifying the details of samples by using microscopes. By shining light onto a sample placed under a carefully chosen set of glass lenses and mirrors, one can see details approaching a micrometre in size (where a micrometre is a millionth of a metre). However, a much higher magnification of more than a thousand times than is obtained in light optics microscopy can be achieved in electron microscopy. Today, the electron microscope is one of the most widely used research tools in understanding materials, in composition and structure, down to an atomic scale. The principle of electron microscopy is to use an energetic beam of electrons to impinge a solid surface where it will either be reflected back or transmitted through the sample being bombarded, depending on the incident electron energy in relationship to the thickness of the solid sample under investigation. The resulting electron signal in this experiment, reflected or transmitted, can be collected and analysed.

The discoveries of Ibn al-Haytham a millennium ago and, in particular, the establishment of the experimental methods based on the study of Optics have triggered an irresistible chain of discoveries over the following centuries, which would reach its culmination in the twentieth century and is still going on today. Majed Chergui reviews the tremendous advances in ultrafast x-ray science, over the past 15 years, making the best use of new ultrashort x-ray sources including tabletop or large-scale facilities. Probing the time evolution of the electronic and nuclear degrees of freedom of these systems on the timescales of femtosecond to picoseconds delivers new insight into our understanding of dynamical matter.

Some of the most important things in life are taken for granted, and nowhere is this truer than with light. Since man first beheld the rising sun in the east, the natural illumination of our world has been the most predictable aspect of the human experience and thus the most apt to be taken for granted. However, in some respects, human evolution can be measured against how we have used technology to transform light from the most passive element of our environment into something active, reactive and even interactive. In the process, light has become more than illumination. For Harry Verhaar, it is important to consider some important milestones marking light's path from human evolution to sustainable revolution, the role Philips Lighting has come to play in this transformation, challenges facing the modern lighting industry as well as the world we live in, and what the future promises. In a sense, a new revolution is at hand.

For Sameen Ahmed Khan, the year-long celebrations of light were a good occasion to revisit the history of optics. The year 2015 had the millennium anniversary of the encyclopaedic seven volumes on optics by the legendary Arab scholar Ibn al-Haytham. A recent analysis by the mathematician and science historian Roshdi Rashed demonstrates that the law of refraction of light had been penned down in detail by Ibn Sahl, who was the teacher of Ibn al-Haytham. In the absence of this major historical discovery, we would have continued to attribute the law to the European Renaissance scientists. Here, we describe the Mediaeval Arab achievements in optics and their influence on the European sciences. Looking ahead, we also briefly examine the ways in which the light sciences can provide a platform for international science collaborations and revival of science in the Arab lands.

9 Photonic Technology

Recent Developments and Challenges of the Twenty-First Century

Azzedine Boudrioua

CONTENTS

Introduction 135
Photonics Technology Issues and Developments 136
  Photonics, a Diffusing Technology 137
  Optical Materials, New Challenges and Issues 138
Optical Fiber as a Vector of Socio-Economic Development: The FFTx Network 139
  Development of the FTTx Network is a Reality 140
Today's Research Produces the Jobs of Tomorrow 140
Conclusion 141
References 141

INTRODUCTION

The International Year of Light and Light-based Technology (IYL 2015) was an important opportunity to celebrate light in all its forms and utilisations. One major area where Light plays a crucial role is optical telecommunications. Light-based technologies that have been developed for optical signal transmission and processing are actually the foundations of any other technologies using Light. The spectacular success of optical telecommunications is primary the result of the boom of the Internet. This development is the fruit of a main effort of research and development in the field of guided optics which led to an improvement in the performances of optical fibres and optoelectronic components able to generate, detect, modulate or commutate light. Consequently, optoelectronic components of any kind at low cost become available in the market, pushing the emergence of other applications in various fields including strategic fields like space and military and also fields of everyday life such as data storage, entertainment, medicine and industry, for instance. In a competing way, the advent of nano-photonics is pushing the limits of photonic devices miniaturisation on scales lower than the wavelength.

The interest in using the photon rather than the electron comes from the very high optical frequencies of the optical signal, which allow a very broad bandwidth and offer an unequalled data transmission capacity.

It is, somehow, amazing to emphasize that all these important developments of Light Science and Technology are mainly based on the principles of light reflection and refraction and their related phenomena. These principles have been the focus of the founding works of Arab scholars during the golden age of Islamic civilization, particularly those of Ibn Sahl and Al-Hasan Ibn al-Haytham (as it is stated in the first part of this book).

Abu Alla Al Sad Ibn Sahl (940–1000), mathematician and author of a treatise on burning instruments around 984 [1]. He explained how lenses and conical instruments deflect and focus light. He analysed parabolic and ellipsoidal burning mirrors. He also examined the hyperbolic mirrors, plane-convex lens and hyperbolic biconvex. His genius idea was to characterise every medium by a constant ratio, a centrepiece allowing him to discover the law of refraction five centuries before Willebrord Snellius (Snell).

Abu Ali al-Hasan Ibn al-Hasan Ibn al-Haytham (died after 1040), the founder of scientific method and modern light science, studied and commented on the works of Aristotle, Euclid, Archimedes and Ptolemy and devoted his life to the study of physics [2]. For him, optics consists of two parts: a theory of vision and physiology of the eye (and the associated psychological perception) and a theory of light with geometrical optics and physical optics. He clarified the difference between the propagation of light and the conditions of vision. He developed a mathematical theory on the model of solid ball movement thrown against an obstacle associated with a geometric approach. For him, the eye is an optical instrument and light is an independent physical entity. This illustrious scientist has really laid the foundations of modern optics including experimental approach.

This contribution gives a highlight of the main concepts and the recent development of photonics technology as well as future challenges.

PHOTONICS TECHNOLOGY ISSUES AND DEVELOPMENTS

Photonics includes all scientific and technological disciplines that are related to Light (photon), its study and its utilisation for the development of functional components and systems. Photonics technologies are increasingly preferred in all sectors of life. We are experiencing a major technological turning point that will let Light be everywhere. This is the third major technological turning point after those of microelectronics and computer science in the 1970s and 1980s, respectively.

Today, Optics is emerging as a promising technology for the next decades. This is mainly due to the maturity of most of the components required for a complete optoelectronics chain and the increased demand for miniaturised photonic circuits at low cost. These needs have led to a multidisciplinary research involving physicists, chemists, engineers and (recently) biologists with the ambition of providing photonics devices with performances built up by considering the matter at the atomic and molecular scales. In addition, as has been the case for the development of microelectronics, optical materials engineering has taken several decades to develop the right materials to produce reliable optoelectronic components. The progress of photonics is undoubtedly determined by the fabrication and characterisation of optical materials capable of being used to generate or transmit light.

PHOTONICS, A DIFFUSING TECHNOLOGY

Light is everywhere: human life exists thanks to Light that we receive from the sun, and nowadays Light is present in almost every technological object used by humankind (Figure 9.1). As previously mentioned, in the field of telecommunications, photonics components play a major role. The development of all-optical networks is likely to increase the presence of optical technologies along the transmission chain. A focus of this field will be the main example of this paper in the next section.

Photonics is increasingly important in life sciences (biology, medicine, etc.). This sector is known as biophotonics [3,4], whose advances concern three main axes: observation of *in vivo* processes at new spatio-temporal scales, diagnosis using new generations of biochips and surgery using invasive phototherapy techniques. Medical instrumentations currently have a wide range of optical devices that are efficient and easy to use thanks to their miniaturisations. For instance, Laser plays a key role in various therapeutic processes. Diagnosis by optical means becomes very important in any public health policy not only for reasons of efficiency but also for economic reasons.

In the space, Photonics Technologies are concentrating on two issues: on the one hand, the integration of optical cables and optronic circuits to reduce the weight of on-board cables (in Satellites or Aircraft) and, on the other hand, the utilisation of electromagnetic insensitivity of photonics circuits. From this point of view,

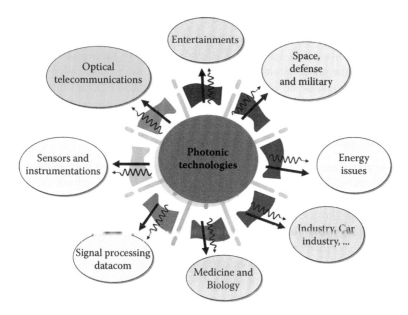

FIGURE 9.1   Photonic technology impacts.

Nanotechnology can also offer very interesting perspectives to miniaturise integrated systems. Optics is increasingly competing with radiofrequency links in several fields of applications. The large bandwidth capabilities offered by light signal can be used for space remote sensing missions and communication between Satellites constellation and between satellites and the ground [5,6]. For example, the first laser transmission and high-speed communication between two Satellites took place in the early 2000s. Projects are also underway for terrestrial-satellite communication using high speed laser transmissions [7].

In the military field, the equipment using light are numerous. For instance, infrared auto-directors, laser-guided shooting systems, night vision systems, imaging and image processing equipment (including holographic analysis of pictures) are some examples of the penetration of light in this strategic sector.

Optics in computers or the 'optical computer' has been one of the motivations behind the development of integrated optics at the early of the seventies [8]. Microelectronics is increasingly facing physical limitations that are difficult to overcome. The problem arises in a crucial way, especially with regard to architectures and interconnections. Today, Optics has key components that can provide alternative solutions to traditional electrical interconnections.

Lighting and display are undergoing important changes due to the utilisation of inorganic and organic light-emitting diodes (LEDs and OLEDs) that should gradually replace conventional incandescent and fluorescent sources [9,10]. Beginning in 2015, the widespread use of white LEDs and OLEDs will save the equivalent of 2 billion barrels of oil, nearly 40,000 MW of electrical power and 50 million tons of CO_2 per year. Analysts predict a 50% reduction in global power consumption for lighting up to 2,025. From the energy point of view, photovoltaic technologies are particularly well suited to the major challenges facing human society: the depletion of fossil resources and climate change. Solar energy is inexhaustible, available all over the world and produces neither waste nor greenhouse gases. This is the reason why the photovoltaic park has developed considerably in the world for nearly 10 years. Indeed, the photovoltaic solar park has grown on average by 35% per year since 1998. It should be noted that the use of nanostructures in photovoltaic solar panels should be able to increase their efficiency.

Optical Materials, New Challenges and Issues

Silicon is very likely the most important material that humankind has ever developed. It is behind the huge development of the microelectronics industry. Extending Si capability to photonics and also merging the two primary device types, photonics and electronics, into one integrated dual functional device will bring the benefits of a large economy of scale and wide application far beyond the simple combination of separate devices. Silicon electronic-photonics circuits are actually the main target of many research and development programs supported by the microelectronics industry [11]. Electronic-photonics convergence is announced for the next generation of computers and nano-photonics prepares integrated circuits of tomorrow.

Another class of materials that will be at the heart of technological revolution is Metamaterials [12]. These are artificial materials with physical properties 'superior' to natural materials. Their properties differ completely from what can be observed in natural materials. More precisely, in the fields of optics and electromagnetism, they exhibit new properties, such as a negative refractive index or an inverted Doppler effect, etc. These structures have some similarities with photonic crystals (PC), already studied for their electromagnetic properties, making it possible to control the propagation of light by photonic bandgap. These PC are rather made up of dielectrics, whereas Metamaterials use metals. Among the fields of research related to Metamaterials are invisibility, super lenses and also plasmonics. This latter is based on the fact that metals exhibit negative permittivity at optical wavelengths related to the resonance of conduction electrons (plasmons). For this reason, it is possible to make resonant structures with dimensions that are much smaller than the wavelength. Negative refraction was demonstrated for the first time by Victor Veselago in 1968, [13] indicating that the optical ray refracted by such material is on the same side of the normal as the incident ray. This amazing property paves the way for an entirely new perspective. Sir John Pendry [14] also showed that these materials with a negative refractive index make it possible to design new optical devices such as a 'superlens' with a theoretically perfect resolution. However, the most fascinating idea of using Metamaterials concerns very likely the invisibility. It is thought that, in the near future, a cloak of invisibility can be achieved by forcing light to bypass an object [15].

Ultimately, the photonics industry is rapidly growing in recent years. It is only at the beginning of its life cycle. Indeed, photonics is experiencing what microelectronics had experienced during the 1970s; but already the impact of these technologies has literally changed many aspects of our daily life.

OPTICAL FIBER AS A VECTOR OF SOCIO-ECONOMIC DEVELOPMENT: THE FFTX NETWORK

Although the telecommunications sector has experienced a slowdown in recent years, similarly to the global economic situation, professionals expect a major development in the next few years. This growth is particularly favoured by the increase in demand for communications, access to the Internet, audio and video transmission, etc.

On the other hand, the development of metropolitan networks and local loops are dramatically increasing [16]. In this context, the optoelectronic components industry is gradually moving towards volume production, with a significant annual price decline (20%–25% in recent years). Similarly, while the active components (sources, detectors and amplifiers) were the most important part of an optical link, the generalisation of Wavelength Division Multiplexing (WDM) and Dense Wavelength Division Multiplexing (DWDM) is also driving more interest to passive components such as filters, multiplexers, compensators and so on. As a result, the current trend is to integrate the maximum number of active and passive functions for reasons of cost and efficiency.

More specifically, the deployment of a high-speed telecommunications network (10, 40 Gb/s and beyond) requires the development of new optoelectronic components and architectures adapted to these transmission speeds. The extension of the usable bandwidth is also an important issue and concerns simultaneously the widening of the C band (C++) L and S bands.

As a whole, in this context, economic development without optical networks can no longer be conceived.

DEVELOPMENT OF THE FTTx NETWORK IS A REALITY

The enormous development of FTTx (Fiber to the x: x may mean Home, Building, Office, etc.) is undoubtedly the most promising sector in recent years. For example, the number of FTTH subscribers is growing around the world. Several suppliers have been offering 1 Gbit/s solutions since many years. The growing importance of FTTA (Fiber to the antenna) driven by 4 G shows an important convergence between optical networks and mobile networks.

For instance, in France, the deployment of the new generation networks at Very High Speed is actually a major industrial and economic task within the framework of the Very Broadband France Plan launched by the government in 2013 [17]. The objective is to deploy the FTTH over the entire country by 2022. This is a major project requiring an investment of €20 billion and offering thousands of jobs. Elsewhere, the trend is the same and some countries have taken a considerable lead. In 2014, the German and Italian governments announced the deployment of very high-speed networks. In January 2015, Barack Obama (United States) emphasised that the deployment of fibre optics in rural areas was a priority at the end of his term.

It should be pointed out that it is virtually impossible to transmit very high bandwidth (>30 Mbits/s) over most existing lines of the local copper loop. This is due to the increased losses as a function of frequency and distance when using copper. For many years now, there has been a gradual transition from copper to optical fibres accelerating the FTTH deployment.

TODAY'S RESEARCH PRODUCES THE JOBS OF TOMORROW

As has been mentioned, Photonics is only at the beginning of its life cycle and many developments are still needed including new concepts and devices. This will need an important investment. However, it should be noted that this is a reliable investment with a real impact on the creation of new jobs and wealth. From the economic point of view, Western countries, notably the USA and Europe, as well as Asia (Japan and China) devote a very large budget to support research and development activities in the field of photonics and nano-photonic technologies. The annual growth of sectors at the European level is estimated at 7.6% over the period 2005–2015 and European production accounts for 19% of world production. Forecasts for the coming years are 1.5 million jobs, 250 billion in revenues and 45,000 patents (Figure 9.2).

Finally, it is important to emphasise that the implementation of these technologies to create a high-tech industrial activity requires an adequate research environment and high-level training in the field of optics and photonics. It also requires the development of a true culture of valorisation, exchange, networking of skills and training through research and for research.

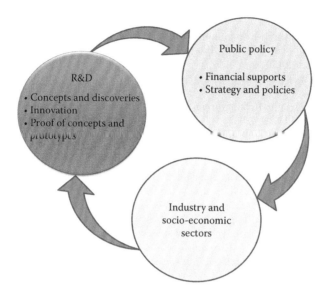

FIGURE 9.2 Today's research produces the jobs of tomorrow.

CONCLUSION

Electronics is gradually giving way to Photonics in various applications ranging from data transmission to medicine and industry. Light is penetrating every aspect of our daily lives ranging from the music we listen to how we communicate with others, our health and security. In fact, we are entering a century that will be full of Light. Today, more than ever, photonics is emerging as a promising technology for the next decades.

Specifically, Arab countries have all the assets (human and material resources) to acquire and develop an advanced photonic technology. Like the work done by the ancestors who, during the golden age of the Islamic civilisation, had undertaken a first phase of translation and understanding of the Greek, Indian and Chinese knowledge before bringing their contributions to human civilisation, it is vital to regain the scientific and technological ground to catch the train of history.

REFERENCES

1. R Rashed, A pioneer in anaclastics – Ibn Sahl on burning mirrors and lenses, *ISIS*, 81, 464–491, doi: 10.1086/355456, 1990.
2. R Rashed, Portraits of science: A polymath in the 10th century, *Science*, 297, 773–773, doi: 10.1126/science.1074591, 2002.
3. P N Prasad, *Introduction to Biophotonics*, John Wiley & Sons, Inc, Hoboken, NJ, 2003.
4. G Keiser, *Biophotonics: Concepts to Applications*, Springer, 2016, ISBN 978-981-10-0945-7.
5. D Cornwell, Space-based laser communications break threshold, *Optics & Photonics News*, 2016.
6. T Tolker-Nielsen and G Oppenhauser, In-orbit test result of an operational optical inter-satellite link between ARTEMIS and SPOT4, SILEX, high-power lasers and applications (International Society for Optics and Photonics), 2002.

7. N Perlot, T Dreischer, C M Weinert and J Perdigues, Optical GEO Feeder Link Design, Future Network & MobileSummit 2012 Conference Proceedings, Paul Cunningham and Miriam Cunningham (Eds.), IIMC International Information Management Corporation, 2012.

8. R Hunsperger, *Integrated Optics: Theory and Technology*, Springer 2009, ISBN 978-0-387-89775-2.

9. Y-S Tyan, Organic light-emitting-diode lighting overview, *Journal of Photonics for Energy*, 1(1), 011009, 2011.

10. M Rodríguez Fernández, E Z Casanova and I G Alonso, Review of display technologies focusing on power consumption, *Sustainability*, 7, 10854–10875, doi: 10.3390/su70810854, 2015.

11. D Thomson, Roadmap on silicon photonics, *Journal of Optics*, 18(7), 2016.

12. G Singh, Rajni and A Marwaha, A review of metamaterials and its applications, *International Journal of Engineering Trends and Technology (IJETT)*, 19(6), 305–310, 2015.

13. V G Veselago, The electrodynamics of substances with simultaneously negative values of ε and μ, *Soviet Physics Uspekhi*, 47, 509–514, 1968.

14. J B Pendry, Negative refraction, *Contemporary Physics*, 45, 191–202, 2004.

15. R Fleury and A Alù, Cloaking and invisibility: A review, *Progress In Electromagnetics Research*, 147, 171–202, 2014.

16. R Ramaswami and K N Sivarajan, *Optical Networks: Practical Perspective*, Morgan Kaufmann (Elsevier), Burlington, MA, ISBN-13: 978-0123740922.

17. http://www.objectif-fibre.fr/

10 Ibn al-Haytham's Thousand-Year Journey from Basra to Mars

*Noureddine Melikechi**

CONTENTS

Introduction 143
Ibn al-Haytham and the Chemistry Camera 144
Perspective 147
Acknowledgements 148
References 148

INTRODUCTION

In the past few years, NASA's Curiosity rover has advanced our understanding of the red planet to an unprecedented level (Grotzinger and Vasavada, 2012). The success of this mission, the Mars Science Laboratory (MSL), is the result of the work of many great scientists and engineers. In this chapter, we propose that more than 1,000 years ago, Ibn al-Haytham's fundamental ideas on optics contributed to scientific discoveries and technological advances that support the Curiosity mission. It is now more and more recognised that Ibn al-Haytham contributed significantly to the birth of the scientific method. Ibn al-Haytham's desire to better understand the world around him, as well as his determination to challenge orthodoxies in place at his time, served humanity. He was the first to suggest that light travels to the eye in rays from different points of an object and can arguably be considered one of the founders of the field of optics.

To determine how science and technology would have evolved had Ibn al-Haytham not published his work is not a trivial task. It is most likely that questions related to the connection between Ibn al-Haytham's full contributions to modern science and technology, including the current work under way on Mars, will remain unanswered for a long time to come. In this chapter, we attempt, somewhat freely, to connect Ibn al-Haytham's ideas to some of the most fundamental scientific aspects that led to the development of one of the instruments on board the Curiosity rover: the Chemistry Camera (ChemCam) (Maurice et al., 2012).

To test the validity of theories and hypothesis, and propose new ones, Ibn al-Haytham observed the world around him and analysed data that he acquired

---

* Member of NASA's Mars Science Laboratory.

143

through experiments. His efforts led to the generation of new knowledge, which through the work of other thinkers and scientists helps us better understand our world and ourselves, excites our imagination and enriches our lives. Today, Ibn al-Haytham's findings are part of our collective knowledge and demonstrate, should it be necessary, that innovation creative thinking and discovery transcend all types of borders, real or perceived, including those related to time and space. Our desire to acquire new knowledge has remained strong throughout history and all across our planet. It is therefore not surprising that the generation of new knowledge depends critically on our ability to relate to one another. This may be best illustrated by Sir Isaac Newton's letter to Robert Hook in which he stated: 'If I have seen further, it is by standing on the shoulders of giants' (Turnbull, 1959).

Humans have long been intrigued by the search of life on planets other than Earth. Owing to its proximity to Earth, Mars has attracted much attention. Life as we know it requires liquid water. Habitability on the Martian surface depends critically on the stable existence of liquid water. Numerous observations of its surface revealed that in its past, Mars *may* have supported the conditions for life to exist (Schiaparelli, 1882; Lowell, 1911). Although there is no direct evidence of the existence of liquid water on the surface of Mars today, recently researchers reported the results of a spectral analysis of seasonally varying features observed in association with equatorial facing slopes. (Ojiha et al., 2015). This study suggests the presence of hydrated salts (sodium, magnesium and perchlorate hydrates) which in solute form can increase the freezing point of water into the range of surface temperatures measured at RSL sites'.

To assess Mars's past and current potential for habitability, NASA launched the MSL mission and successfully landed the Curiosity rover on the planet in 2012 (Grotzinger et al., 2012). On board the Curiosity rover, in addition to ChemCam, there are nine research instruments designed to survey the performance of Curiosity during its descent, evaluate its surroundings on Mars, and investigate samples of Martian rocks, soils and atmosphere.

IBN AL-HAYTHAM AND THE CHEMISTRY CAMERA

ChemCam relies on our knowledge of atomic and molecular physics as well as fundamental and applied optics. Names that are synonymous with major contributions to atomic, molecular and optical physics include giants such as Descartes, Newton, Huygens, Maxwell, Fermat, Einstein and others. However, even a casual inspection of the present book reveals the breadth and depth of intellect and the creativity of a man who has made significant contributions to the birth of the modern scientific method and to the field of optics: Ibn al-Haytham. In his book, *Kitab al-Manazir*, translated into Latin as *Opticae thesaurus Alhazeni* in 1270, Ibn al-Haytham gave the first correct explanation of vision, showing that light is reflected from an object to the eye and not vice versa as was understood to be the case by his predecessors. Before him, both Ptolemy and Aristotle suggested that light was emitted either from the eye to illuminate objects or generated from objects themselves.

ChemCam consists of a laser-induced breakdown spectroscopy (LIBS) set-up and a remote micro-imager (RMI) system (Maurice et al., 2012). LIBS is a spectrochemical analytical technique that can provide without preparation the elemental

Plasma plume

FIGURE 10.1   Image of a plasma plume formed by shining an infrared laser light (not seen) on a plate of metallic object. Both the core (yellow) and cold zone (blue) show visible colors of light emitted. This image illustrates Ibn-al-Haytham statement: the yellow and blue lights seen are emitted as the result of 'strucking' the 'object' with a primary light source, in this case an infrared laser.

composition of a sample in a solid, liquid or gaseous phase (Miziolek et al., 2008; Cremers and Radziemski, 2013). The technique is based on focusing a pulsed laser with irradiance large enough to generate thermal energy in the target, which yields vaporisation and ablation. As the ablated part expands, the vapour continues to absorb optical energy from the laser and becomes a plasma plume. This plume, illustrated in Figure 10.1, collides with the much cooler environment that surrounds it and begins to cool down, emitting optical radiation through various emission and recombination processes. This radiation is collected and spectrally analysed to obtain information on the target. Interestingly enough, more than 1,000 years ago, Ibn al-Haytham observed that objects struck with light can become luminous and radiate secondary light. Such an insight is a testament to his level of observation, analysis and scientific intuition and his willingness to explore novel ideas. With no knowledge of the fundamental building blocks of matter, he was able to observe that under conditions, when objects interacted with light, 'secondary' light could be emitted. To arrive to such a conclusion reveals a sophisticated level of creativity and a deep interest for studying light-matter interactions in a way that was very revolutionary during Ibn al-Haytham's time and indeed for many years after.

In addition to LIBS, ChemCam consists of the RMI camera system, which captures detailed images of potential ChemCam targets, such as Martian rocks and soils, as well as context images of these targets (Maurice et al., 2012). Ibn al-Haytham contributed to the better understanding of the relationship between light and human vision. He gave the first correct basic explanation of vision, showing that light is reflected from an object to the eye and made fundamental contributions to the study of binocular vision. Applications of stereo vision include virtual reality and visualisation of volumetric medical images such as magnetic resonance imaging and computerised tomography. In contemporary stereovision, two cameras are used to obtain two differing views of a scene that resembles the human binocular vision. Assuming

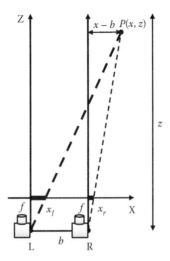

FIGURE 10.2 Depth perception from stereo.

that two cameras or eyes, each associated with a lens of focal length f, are aligned so their X-axes are collinear and the Y-axis and Z-axis are parallel, we can calculate the location of 3D points by using simple geometric models (Figure 10.1) and solving the following simple equations (Figure 10.2):

$$z = \frac{fb}{x_l - x_r} = \frac{fb}{d}, \quad x = \frac{x_l z}{f} = b + \frac{x_r z}{f}, \quad y = \frac{y_l z}{f} = \frac{y_r z}{f}$$

Thus, the concept behind the design and functioning of the ChemCam camera and other contemporary cameras is based in part on Ibn al-Haytham's contributions to ray optics and binocular vision.

In addition, Ibn al-Haytham, in *Kitab al-Manazir*, investigated the reflection of light from spherical, parabolic and cylindrical surfaces irrespective of its source, for example, sunlight, light from fire or light reflected from various geometrical surfaces.

ChemCam acquires remote laser induced breakdown spectra at distances that vary between 1.6 and 7 m from the instrument itself. The variation in the distance between the instrument and the target selected for analysis is a major source of uncertainty for the measurement because the intensity of the light emitted by the plasma decreases with distance in non-trivial ways (Melikechi et al., 2014; Mezzacappa et al., 2016). Thus, to compare two or more spectra acquired at different distances with enhanced accuracy – and therefore to further our understanding of the geochemistry of the red planet – it is necessary to take into account distance effects. In this context, Ibn al-Haytham was the first to show, through measurement, that light loses its intensity as it propagates (Rashed, 2015).

Ibn al-Haytham's work on optical reflection led him to suggest 'that the capacity of a surface to reflect light was due to the close packing of the material surface'.

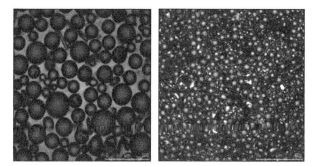

FIGURE 10.3 Random packing of disks (stones) with two different sizes: The reflected light from the smaller disks (right) will be more specular than diffuse than the one originating from the larger disks (left). One can imagine that Ibn al-Haytham did observe that the nature of light emitted from various objects varies with the structure of the object itself.

(Bettany, 1995). This observation suggests that Ibn al-Haytham realised the importance and role of the nature of the surface with which light interacts in determining the properties of the light reflected from it. From this point, one can imagine that Ibn al-Haytham could have connected the nature of the image formed after reflection with the structure of the surface of the object. In addition, Ibn al-Haytham noted that light can be refracted by going through partially transparent objects, as well as reflected by striking smooth objects such as mirrors, travelling in straight lines in both cases. Ibn al-Haytham did not refer to the types of light that emerge from various surfaces as specular or diffuse. However, he did realize that the structure of the reflecting surface and the 'quality' of the image formed are somewhat linked. Today, packing of a material, defined as the fraction of the total volume of the system occupied by solids (the *volume fraction* of the system), is known to be an important physical property of a multiparticle solid system. As the particle size distribution increases, the particles pack to higher volume fractions because the smaller particles pack more efficiently by either layering against larger particles or by fitting into voids created between neighbouring large particles (Brouwers, 2006; Desmond and Weeks, 2014, Sivakumar et al., 2014). This is illustrated in Figure 10.3.

PERSPECTIVE

Has Ibn al-Haytham's work influenced the evolution of ideas on the properties of reflected light from various objects? Has his work influenced that of Erasmus Bartholinus (1625–1698), who in 1669 discovered the double refraction effect using a variety of calcite samples, or that of Christian Huygens (1629–1695) who investigated double refraction? Was Ibn al-Haytham himself influenced by the work of the Vikings who investigated skylight for navigational purposes? Has Ibn al-Haytham's work impacted the largest science mission to Mars? Through his quest for truth using a methodology based on measurements and his significant contributions to optics, Ibn al-Haytham has set a strong foundation for many areas of science and technology including those focused on space exploration such as the Mars Science

Laboratory mission. Ibn al-Haytham has been able to do his work in part because he did challenge the preconceived ideas of his time. For centuries, the Muslim civilisation produced people who could and did ask questions that excited their imagination. As a result, it gave birth to some of the greatest minds in mathematics, physics, astronomy, chemistry, philosophy and medicine. Ibn al-Haytham and other Muslim scholars, like other great minds from all over the world, have contributed tremendously to humanity's quest for a better understanding of our world. Their work remains a testimony to what humans can achieve when and wherever challenging orthodoxies is not discouraged or only tolerated. Today, the West offers a richer environment for the pursuit of scientific inquiry. It attracts scholars from around the world whose dream is to make new discoveries, develop new technologies, better understand societies and contribute to their well being. Sadly, relations between the Muslim world and the West are often dominated by misunderstanding and mistrust. Building or rebuilding this trust will be substantially enhanced as stronger bridges with science as their foundation are opened or reopened. The future of our planet and of humanity will be brighter. It is time to challenge ourselves to do our best to serve Humanity irrespective of where we are and irrespective of the challenges we face. More than 1,000 years ago, Ibn al-Haytham has shown that this is possible. Surely, we can at least try.

ACKNOWLEDGEMENTS

The author thanks Amir Khan, Hacene Boukari, Renu Tripathi, Sokratis Makrogianis, Gour Pati and Omar Melikechi for their helpful comments.

REFERENCES

L Bettany, Ibn al-Haytham: An answer to multicultural science teaching? *Phys. Education*, 30, 247, 1995.

H J H Brouwers, Particle-size distribution and packing fraction of geometric random packings, *Phys. Rev. E.*, 74, 031309, 2006.

D E Cremers and J J Radziemski, *Handbook of Laser-Indiced Breakdown Spectroscopy*, Wiley, 2nd edition, 2013.

K W Desmond and E R Weeks, Influence of particle size distribution on random close packing of spheres, *Phys. Rev. E.*, 90, 022204, 2014. DOI: https://doi.org/10.1103/PhysRevE.90.022204

J Grotzinger and A Vasavada, Reading the red planet, *Sci. Am.*, 307, 40–43, 2012.

J P Grotzinger et al., Mars science laboratory mission and science investigation, *Space Sci. Rev.*, 170(1), 5–56, 2012.

P Lowell, *Mars and Its Canals*, The Macmillan Company, New York, p. 453, 1911.

S Maurice et al., The ChemCam Instrument Suite on the Mars Science Laboratory (MSL) Rover: Science objectives and mast unit description, *Rev. of Space Inst.*, 170(1), 95–166, 2012.

N Melikechi et al., Correcting for variable laser-target distances of LIBS measurements with ChemCam using emission lines of Martian dust spectra, *Spectrochimica Acta Part B Atomic Spectroscopy*, 96, 51–60, 2014.

A Mezzacappa et al. and the MSL Science Team, Application of distance correction to ChemCam measurements, *Spectrochimica Acta Part B Atomic Spectroscopy*, 120, 19–29, 2016 and 115, 2011/2015, DOI: 10.1016/j.sab.2015.11.002.

A W Miziolek, V Palleschi and I Schechter, *Laser Induced Breakdown Spectroscopy; Fundamentals and Applications*, Cambridge University Press, 2nd edition, 2008.

L Ojiha et al., Spectral evidence for hydrated salts in recurring slope linea on mars. *Nat Geosci.*, 8, 829–832, 2015.

R Rashed, Plenary Lecture, UNESCO's Year of Light Celebration, Paris, France, September 16, 2015.

G Schiaparelli, On some observations of Saturn and Mars, *Observatory*, 5, 221–224, 1882.

P Sivakumar, L Taleh, Y Markushin and N Melikechi, Packing density effects on the fluctuations of the emission lines in laser-induced breakdown spectroscopy, *Spectrochimica Acta Part B-Atomic Spectroscopy*, 92, 84–89, 2014.

H W Turnbull, ed. *The Correspondence of Isaac Newton: 1661–1675*, Volume 1, London, UK, Published for the Royal Society at the University Press, p. 416, 1959.

11 Ibn al-Haytham and the International Year of Light
His Legacy

Mohamed M El-Gomati

CONTENTS

Study of Light Optics 151
Al-Hassan Ibn al-Hassan Ibn al-Haytham (965–1040) 152
Was Ibn al-Haytham imprisoned? 153
Use of Electron Microscopy as a Fundamental Research Tool 154
Spherical Aberration 155
In Pursuit of Efficient Light, the Blue LED and Electron Microscopy 158
Ibn al-Haytham and the Scientific Method of Enquiry 160
Ibn al-Haytham and the Camera Obscura: Whose Camera is it? 161
Acknowledgements 163
References 163

Light has fascinated humans since time immemorial. While we see huge advances today in relation to light and light-based technologies, one needs to remember that our advancements and understanding of light and optics are built upon accumulated contributions from all corners of the earth. In highlighting the importance of light in our lives today, it is useful to recall how some of the early contributions in understanding the role of light in our lives, for example, how we see, have come to pave the way towards later developments, inventions, methodologies and understanding in science and developments in technology that we still use today.

Our regular dependence on light for our daily needs has thus far been so vital to our lives that we take it for granted. The early Egyptians realised the importance of light in their buildings of the pyramids whose shapes were built to reflect the sun's rays. The Chinese, the Indians as well as the South and Central Americans have had their developments of light related objects, too.

STUDY OF LIGHT OPTICS

However, the study of light optics as a science of its own formulating its properties was sparked by Arab scholars questioning the Greek legend which claims that

151

Archimedes' troops used the sun's rays reflected off mirrors to set aflame the Roman ships attacking Syracuse in 213 BC. During the ninth and tenth centuries CE, a number of Arab scholars (among them are e.g., Abu Ishaq al-Kindi (d873), Qusta Ibn Luqa (d912) and Ibn Sahl (d1000)) who read this claim got interested in studying it through understanding the optical properties of mirrors and glass lenses.

The most distinguished and prolific of these early Arab authors on optics, however, and without a doubt was al-Hassan Ibn al-Hassan Ibn al-Haytham, known in the west by his Latinised name Alhacen. It would be beneficial to mention information about this outstanding scholar here, as well as to dispel some of the myths that surround his story and legacy. Readers are encouraged to consult the work of several historians of science, most notably, the writing by Roshdi Rashed (see, e.g., a recent publication, Ibn al-Haytham and analytical mathematics, Routledge, 2013) who dedicated his lifetime works to the subject. The aim of this article is to introduce the reader to developments made by early Arab scholars, exemplified in the work of Ibn al-Haytham in his work on optics, which we still use today.

AL-HASSAN IBN AL-HASSAN IBN AL-HAYTHAM (965–1040)

Who was Ibn al-Haytham? What is his correct name? Where was he born and where did he spend his adult life? What are his major contributions in optics and, more widely, science? Detailed research by Roshdi Rashed (1996, 2013) as well as many earlier publications, notably Nazif (1942) and Sabra (1983) clarifies some of the confusion that has found its way into published material on this Arab scholar. The first of these concerns his name which he (Ibn al-Haytham himself) signed as *al-Hassan Ibn al-Hassan Ibn al-Haytham*. He was born in Basra (Iraq, 965) and lived most of his adult life and died in Cairo (Egypt, 1040). It is often found that writers, biographers and bibliographers mix and mistakenly attribute to Ibn al-Haytham works by another scholar, Mohammad Ibn al-Hassan, who lived in Baghdad (Iraq) around the same time as Ibn al-Haytham. The confusion to some might have arisen because of the name *Ibn al-Hassan,* but careful biographers should not make such trivial mistakes. Mohammed Ibn al-Hassan wrote mainly on mathematics and philosophy while Ibn al-Haytham wrote on mathematics as well as optics, astronomy, philosophy and medicine.

Interesting too are some references to Ibn al-Haytham as al-Misri (i.e., the Egyptian). A person can sometimes be associated with their city or country of birth, so it is possible to refer to Ibn al-Haytham as 'al-Basri', referring to his birthplace, the city of Basra in Iraq. It is perhaps easy to see how such confusion could arise in this case (both in English and Arabic) by exchanging the letter b̲ for the letter m̲, so that his name will be referenced as al-*M̲asri* instead of al-*B̲asri*, hence the reported confusion about his name.

It is most likely too that he travelled to Egypt in pursuit of a major project to regulate the flooding of the river Nile, which he was unable to realise. This failure to fulfil such a project would have been a disappointment to the Egyptian ruler at the time who greatly desired it. The ruler was also known for his ill temper, if not brutality, in dealing with his countrymen, and such a failure could quite conceivably have resulted in the meting out of punishment to Ibn al-Haytham. What is clearly disputed by Rashed in his comprehensive study, however, is the imprisonment of Ibn al-Haytham by the Egyptian ruler.

WAS IBN AL-HAYTHAM IMPRISONED?

The major biographers of Ibn al-Haytham, including Ibn Abi Usaibi'a (1965), Abū al-Ḥasan 'Alī ibn Zayd al-Bayhaqī (1976) and al-Qifti (1903), refer to the imprisonment of Ibn al-Haytham on his return from the upper Nile upon declaring that he could not realise his idea. They further go on to claim that Ibn al-Haytham had to feign madness in order to escape the wrath of the Egyptian ruler for his failure; a fate of death would almost certainly have been the result. However, in the Islamic tradition, as well as in many others, insanity would spare one from the death penalty.

One area that is not normally discussed in many of these writings claiming the imprisonment of Ibn al-Haytham is his scientific writing during this period. Had Ibn al-Haytham been imprisoned, he would certainly have been unable to write the books which are known that he published during the years 1010 and 1020, which is the period he was allegedly locked-up in prison and most likely unable to write (see list of books written by Ibn al-Haytham and compiled in Rashed, 2013, pp. 392–423). However, it is also important to pose the following question: was Ibn al-Haytham a totally free man during this period instead of being imprisoned and totally in isolation of the outside world as a result of his failure to build a structure to regulate the flow of the Nile? Taking into account the ruler's annoyance with Ibn al-Haytham's failure with a project that the ruler wanted to realise, it is also unlikely that Ibn al-Haytham would have been a totally free man either! So if Ibn al-Haytham was not imprisoned, could an alternative explanation of the details surrounding this period be a milder form of imprisonment? Could the result of a fallout between him and the Egyptian ruler, who has certainly been referred to as the mad Caliph, as well as being known for his violent temper and whose life ended in an assassination in the year 1020, have been that Ibn al-Haytham was put under restricted movement, equivalent to a modern-day 'house arrest'? Such restricted movement would allow Ibn al-Haytham some interaction with the outside world, albeit limited, but sufficient enough for him to pursue his scientific writings, as was evident from the dates of some of the manuscripts written by him (R Rashed, 2016, private communications). This, to me is a more likely scenario, as he managed to write several manuscripts between the year 1010, when he arrived in Egypt, and the year 1020 when the Egyptian ruler died.

Whatever the truth is about Ibn al-Haytham's interaction with the ruler of Egypt and the many stories surrounding this interaction, this scholar wrote a large number of manuscripts, estimated to be 96 in total, and which have been traced back to him as their author by dedicated scholars like Rashed and Sabra, to name two of the most knowledgeable researchers of the writings of our scholar.

To mark the year 2015 as the year of light and light related technology, this cannot be complete without visiting the contributions of Arab scholars in this area of science, and with Ibn al-Haytham being an individual who particularly played a pivotal role in the development of this branch of science. However, did Ibn al-Haytham flourish on his own, only relying on Greek translations? Or was there a thriving community of Arab scholars who preceded him and whose work was also important to the development of optics as a science of its own?

In what follows, I will give only a small number of examples, which are related to one area of research I am involved in, due to the limitations on the length of the present article, as well as the theme of the IYL. I will try to relate some of the contributions of early Arab scholars in optics, to some of the current state-of-the-art research and technology that is also related to optics. In this context, I will use the pursuit of scientists and engineers in developing more efficient and environmentally friendly light sources to the less conventional incandescent light bulbs as an example to illustrate this relationship.

USE OF ELECTRON MICROSCOPY AS A FUNDAMENTAL RESEARCH TOOL

Microscopy, the art of magnifying the details of samples, has been a great help to humanity. By shining light onto a sample placed under a carefully chosen set of glass lenses and mirrors, one can see details approaching a micrometre in size (where a micrometre is a millionth of a metre). However, a much higher magnification of more than a thousand times than is obtained in light optics microscopy can be achieved in electron microscopy. Today, the electron microscope is one of the most widely used research tools in understanding materials, in composition and structure, down to an atomic scale. The principle of electron microscopy is to use an energetic beam of electrons to impinge a solid surface where it will either be reflected back or transmitted through the sample being bombarded, depending on the incident electron energy in relationship to the thickness of the solid sample under investigation. The resulting electron signal in this experiment, reflected or transmitted, can be collected and analysed. These two geometries of electron–solid interaction have resulted in two types of instruments known as the scanning electron microscope (SEM), which normally uses one of the reflected electron signals from the solid surface, while the other type uses the transmitted electron signal and is referred to as the transmission electron microscope (TEM).

Figure 11.1 depicts the geometry of an experiment in the reflected mode, and lists some of the most widely used signals to form an image that reflects some property of the solid under investigation. If the sample under electron bombardment in this figure is thin enough, in comparison to the energy of the incident electrons such that it allows these impinging electrons to transmit through, then it is possible to collect the transmitted electrons which would reflect a different but complementary signal to those collected in the reflection geometry. These detectors range from ones depicting the atomic arrangements of the sample, to others indicating the sample's elemental composition and sometimes its chemical state. Figure 11.2 depicts schematics of possible configurations of both instruments, which come in a variety of sizes. While the majority of these require a dedicated space of several square metres, as is often the case of TEMs up to 20 or 30 m^2, the latest trend, particularly in SEMs, is the development of desktop type use occupying less than 1 m^2.

It is interesting to note that the laws of physics governing the path of the electrons through the various lens components in electron microscopy share a great deal of

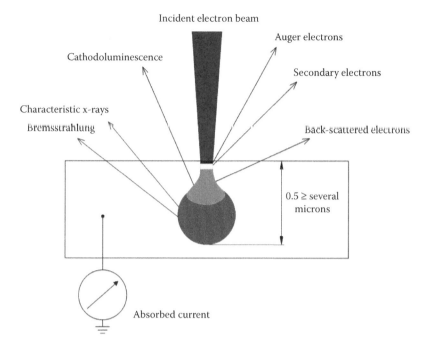

FIGURE 11.1 Schematic illustration of electron–solid interaction in the reflected mode and where the sample thickness exceeds the electron range in the solid. The diagram also shows the various signals that result from such interaction.

similarities with its ray optical counterpart, that is, optical microscopes. Two important properties of optical and electron lenses refer to deflection of the incident light ray (or electron) crossing a boundary between two media of different refractive indices (or fields for electrons) which acts to alter the light or electron's direction of motion, respectively. This relationship is referred to as Snell's law of refraction. The same relationship was also arrived at in a study of optical lenses in the second part of the tenth century by an Arab scientist by the name of abu al-Alaa Ibn Sahl, who lived in Baghdad, today's Iraq. The discovery of the manuscript describing this development was made by Rashed (1990). Figure 11.3 depicts a representation of Ibn Sahl's demonstration of the law of refraction which he made in his studies of optical lenses.

SPHERICAL ABERRATION

Another property, which is both central and crucial to the use and operation of an electron microscope at the highest possible resolution, (i.e., the ability to resolve the details of two closely adjacent features in a sample being investigated), is the minimisation of the spherical aberration effect. The origin of this effect stems from the use of circular apertures in the microscopes (both in ray optics or in electron optics), to reduce the diameter of the electron beam being generated, while at the same time using another aperture to limit the number of electrons being used to form the electron beam. Spherical aberration also applies to mirrors, as will be shown later on.

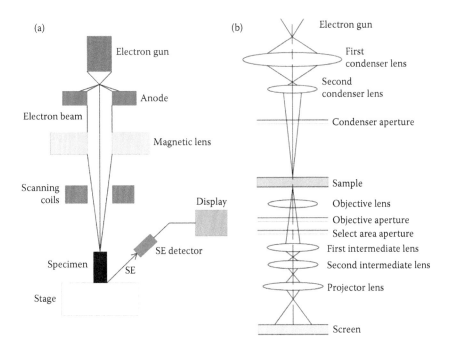

FIGURE 11.2   Schematic illustration of the various components used to form images in (a) the SEM and (b) the TEM, respectively.

Figure 11.4 is an illustration of the origin of the spherical aberration in round lenses. In the present case, it is easily shown that incident electrons passing through a lens will focus at different points along the central line of the lens. In order to practically use the incident electrons to image, one needs to minimise the size of their focal point and this is normally taken as the disc of least confusion (DLC),

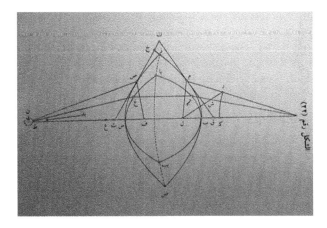

FIGURE 11.3   Ibn Sahl's law of refraction, also known as Snell's law of refraction. (Adapted from Siecle Ibn Sahl, Quhi et Ibn Al-Haytham, Geomtire et Dioptrique au X', 1996.)

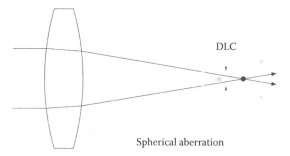

FIGURE 11.4   Schematic diagram illustrating Spherical aberrations and the definition of the disc of least confusion (DLC).

as depicted in the figure (i.e., the smallest diameter of the electrons' trajectories which appear to be focusing along the central axis). However, the extent of this focal distance away from the ideal focal point of such a lens is a function of the angles subtended between that focal plane and the diameter of the beam defining aperture. This in turn means that the larger the aperture hole diameter, the greater the effect of the spherical aberration on the size of the electron beam.

The effect of spherical aberration was first noted and correctly identified by the Arab scientist Ibn al-Haytham (d1040) in his masterly work on Optics, 'Kitab al-Manazir'. While this defect was known early on during the electron microscopy development in the early 1930s, an effective solution was only proposed and realised in 1997 (Krvanek et al., 1997). Again, the effect of spherical aberration is more obvious in the TEM case, which, when corrected for spherical aberration, operates at a resolution often of the order of 0.1 nm or less as shown in modern instruments. Its correction in the SEM is also equally important but is only applicable in top-of-the-range instruments. Figures 11.5 and 11.6 shows the effect of correcting a spherical aberration in a modern TEM (C Humphreys, 2016, private communications).

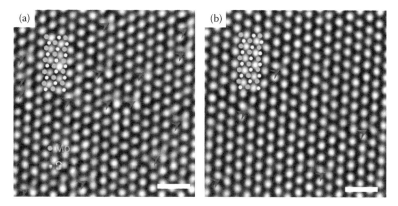

FIGURE 11.5   Effect of correcting spherical aberration in a TEM (a) before and (b) after. Note that the red arrows refer to points where the separation between the imaged atoms becomes clearer after the spherical aberration correction. (Adapted from Zhihao, Y et al. 2014, *Nat. Commun.*; 5: 5290, doi: 10.1038/ncomms6290.)

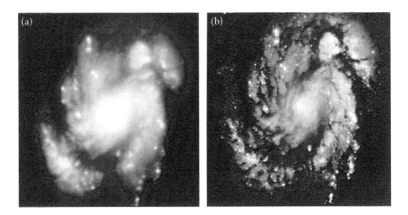

FIGURE 11.6 Hubble Space Telescope images (a) before and (b) after the STS-61 mission. (Adapted from Hubble telescope mission, http://www.nasa.gov/mission_pages/hubble/servicing/index.html)

It is also equally important to note that the spherical aberration effect is inherent in all cylindrical or spherical lenses, such as cameras and telescopes, which if not corrected would compromise the quality of the resulting images. This is what happened with the first images collected by the Hubble space telescope (Hubble telescope mission, 1990), a project that cost several billion U.S. dollars and took several years to complete. The first images collected were of very poor quality as shown in Figure 11.6a. The fault causing such a poor image was identified as the spherical aberrations of one of the telescope's mirrors and a repair mission was launched to fix it. This consisted of a number of astronauts, one of whom was an experienced experimental scientist, who was able to walk in space to repair the fault. Figure 11.6b shows the effect of correcting the spherical aberration on the quality of the images obtained.

IN PURSUIT OF EFFICIENT LIGHT, THE BLUE LED AND ELECTRON MICROSCOPY

While we celebrate light and light related technology, we often take for granted the sun as a source of 'natural light' and energy, both abundant and free! Mankind has, over the years, developed light sources that address these two points: as close as possible to natural light while also using the least amount of energy to produce. The incandescent light bulb, still by and large the most widely used light source, is also without a doubt the least efficient in both accounts. Other forms of light sources have been developed but none has equalled the benefits of sunlight, particularly on a health basis, where, for example, sunlight is the main source for replenishing vitamin D in humans by about 90% against only 10% via food (Dowed, 2012). Natural light is also believed to have a protective effect against certain types of cancers, breast and prostate types, by preventing the overproduction of cells (Public Health England Report, 2007). A new idea towards realising a light source resembling natural light is based on a feature of one of the most basic semiconductor electron devices, the p–n diode. It is estimated that if such a new source replaced the conventional ones in the United States in homes and offices, a saving of approximately \$30b per annum could result, in addition to about 1,800 metric tonnes of carbon emissions (report by the USA Department of Energy, 2014).

A p–n diode is a device made out of two semiconductor materials. One part, referred to as the n-type, is characterised by being rich in electrons, while the other type is characterised as being rich in holes, referred to as p-type material (note that a hole is essentially an electron vacancy). In the most widely used p–n diodes, the materials used are made out of silicon which starts with an equal amount of electrons and holes. To convert it to an active device, a minute amount of another material rich in electrons or holes is added to convert it to either an n- or p-type material. When these two different type of materials are brought into contact with each other, electrons and holes combine (i.e., an electron fills in a hole vacancy) and give out some energy that can be captured in the form of light. In this case, the diode is referred to as a *light-emitting diode* (LED). This emitted LED light has not always been the sought after sun resembling 'white light' as it comes in different colours depending on the materials used. It was proposed that if a number of different LED coloured lights, red, green and blue, could be produced, then when these are combined, they would yield a white light. Red and green LEDs have been developed for over 50 years, but the missing colour, the blue LED needed to produce the promised sun resembling 'white light', has eluded researchers for many years.

The search for the blue LED component of such a combination was extensively studied during the 1980s and 1990s. The two research groups, from Japan and the United States, which contributed to the successful realisation of such a source were deservedly awarded the Nobel Prize for Physics in 2014 (compare with the discussion about the camera obscura and the acknowledgement of its developers in this article). The materials finally used to produce the blue LED comprised a combination of gallium and nitrogen, called gallium nitride (GaN). However, it was with the aid of electron microscopy that this material in the right composition and atomic structure was finally developed. Figure 11.7 shows the application of spherical aberration correction on TEM obtained images, and the identification of the atomic interface between silicon and aluminium nitride (C Humphreys, 2016, private communications).

It should be mentioned here that while only two examples have been given to show the use of early developments in optics by early Arab scholars, the field of electron microscopy continues to benefit from the use of optical laws developed by early scholars like Ibn Sahl and Ibn al-Haytham in pursuit of new frontiers in science. A couple of these developments are briefly referred to here. One concerns the use of much lower incident electron energies in comparison to electron energies used in conventional SEMs, in what has become known as scanning low energy electron microscopy. The author has been active in the development of this technique and in the interpretation of the results obtained, which are sometimes markedly different and surprising when higher energies are used (El-Gomati et al., 2004, 2005). In addition, a trend in reducing the overall size of the electron microscope column, which will be of advantage in reducing the footprint of the instrument as well as in using it in new ways (such as space missions, etc.), has also begun (Spallas et al., 2006).

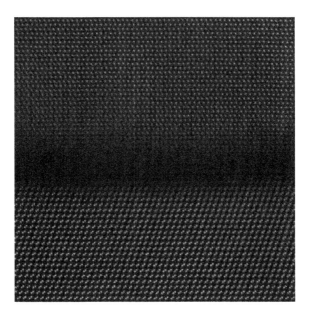

FIGURE 11.7 Showing a resolution of 0.1 nm obtained for imaging the growth of AlN on top of an Si<110> single crystal. The image clearly shows a boundary where atoms are not arranged in a systematic manner as is normally found in single crystals. The interfacial area is composed of an SiN layer with a random distribution of the atoms, which is referred to as amorphous material.

The other recent development in electron microscopy concerns releasing the incident electron beam to impinge a sample in pulses at the rate of femto seconds (i.e., instead of continuous electron bombardment as in conventional electron microscopes. Note a femto second is 10^{-15} sec). This novel configuration enables the researcher of the following dynamic events resulting from the electron–solid interaction as a function of time (Zewail and Thomas, 2009).

IBN AL-HAYTHAM AND THE SCIENTIFIC METHOD OF ENQUIRY

The example given above for the development of white LEDs was only realised because the researchers pursuing it followed what we refer to as the scientific method of enquiry. This method is also generally followed in all applied research in science and engineering. Although it is often claimed that this method is due to the founders of the Royal Society (London, UK), one should acknowledge that it was in fact Ibn al-Haytham who laid down the tenets of this method. In so doing, Ibn al-Haytham used an Arabic word 'al-mutabir' to refer to the person carrying out the experiment. The verb *i'tabara*, denoting experimenting, was also frequently used in his writings and defined his methodology in research. While this term has also been used by earlier and contemporary Arab scholars to Ibn al-Haytham, what he (Ibn al-Haytham) gave in his usage of this term was a deeper meaning than just what the word mean in Arabic, that is, to carry out an experiment without first postulating what the experiment is all about, carrying it out and finally analysing the results against the postulates. What he demonstrated with this term was its use for 'the examination of things and their signs in order to deduce what is latent on the basis of the manifest' (Rashed, 2005).

IBN AL-HAYTHAM AND THE CAMERA OBSCURA: WHOSE CAMERA IS IT?

There has been a great deal of discussion over the years regarding the invention of the camera obscura as an imaging tool. Was it a Chinese invention? Or was it an Arab-Muslim contribution during the Muslim golden age of science? Or was it much later than either of these two claims? Who did what, when and why remains a confused and unanswered question in many recent writings.

Firstly, an important point that needs to be clearly stated for enriching the discussion around this topic, and consulting and relying on a careful reading of 'Kitab al-Manazir', the book of optics by Ibn al-Haytham, is that the pinhole camera, which was first translated into Latin as the 'camera obscura' as suggested by Ibn al-Haytham, was not meant to be an imaging tool. What Ibn al-Haytham suggested is an experimental set-up he called 'albit al-muzlim', which translates into Latin as the 'camera obscura', to aid his fundamental studies of ray optics. He did so and reported his findings in his book of optics, 'Kitab al-Manazir'. It is also probably true that in the fifth century BC the Chinese scholar Mo-Ti discovered the imaging properties of the pinhole camera, but to me this remains an observation rather than an invention of an imaging tool. More important, however, is that no reports exist to link Ibn al-Haytham to Chinese science whatsoever. So even if it is true that the Chinese scholar Mo-Ti discovered such a property, there is no evidence that this was transmitted outside of China. In addition, the Mo-Ti work, which is the only written work we know of discussing the action of the pin-hole camera in China, this writing is only a recent discovery by historians interested in the Chinese history of science, and therefore confirms that there was no direct or indirect link between Mo-Ti's writing and Ibn al-Haytham's pioneering work. In other words, there is no direct link between Chinese science and particularly the Camera obscura and the western developments of this device. On the other hand, there is direct linkage between scholars of the renaissance and the work of Ibn al-Haytham which was translated into Latin more than once (Smith, 2001).

Secondly, there is also some confusion in the Arabic literature regarding the word 'camera' and its relationship to the word 'qamra' which is used to refer to a private room. In Arabic, one would say, when describing the cockpit in an areophane: 'qamrat ul-qiada', meaning the private room the pilots use for flying the areophane. In English, people would use the word 'camera' to refer to privacy, such as 'the court is in camera', meaning the court is in private deliberation. To the best of my knowledge, I could not find a root for the word 'qamra' in Arabic to mean privacy, and modern Arab dictionaries refer to it as an Arabisation of the word 'camera'; both meaning a private, if not a darkened, room.

Thirdly, equally if not more important in this debate is the research finding and careful studies by Professor Rashed regarding Ibn al-Haytham's work on burning spheres and the imaging characteristics of glass spheres. These studies have revealed to us that Ibn al-Haytham understood the imaging properties of the glass sphere and in so doing discovered and correctly identified an important factor that blurs the image, which he called 'spherical aberration'. It is the same spherical aberration we refer to today in mirrors and lenses (see discussion above on telescopes and microscopes, whether ray or electron optics type); that is, more accurately, it is the same spherical aberration property we struggle to reduce its effects in all imaging devices, such as lenses, mirrors, cameras and microscopes.

As for the camera obscura and its relationship to Ibn al-Haytham, it would have been almost impossible if Ibn al-Haytham did not see an image of the candles he used in his dark room, which he called al-Beit al-Muzlim and we now call camera obscura, thanks to the Latin translation of 'dark room', to ascertain that light rays do not mix up as they cross each other. It is almost inconceivable that a person of such an ability and relevant experience in this subject, and one who was investigating the formation of images in the eye, would not have realised and almost certainly seen the important effect the size of the hole in his dark room, *aca al-Beit al-Muzlim*, have on the quality of the obtained image of the candle flames that are used as a source of light being transmitted through the small hole in front of the said candles. In other words, Ibn al-Haytham must have realised the relationship between the size of the hole and the quality of the image obtained, which is an effect of the spherical aberration showing that if the hole size is enlarged, then the image becomes blur. That person would surely, as a brilliant experimentalist, have seen that a relatively small hole that allows sufficient light to be transmitted and projected on the wall opposite the said hole shows a crisp image of the candles. He/she would also have seen that the larger the diameter of the hole, although it may improve the brightness of the projected image of the candle flames, does indeed compromise its sharpness (i.e., makes it blur), due to nothing other than the effect of spherical aberration. This relationship of the imaging property of 'al-Bit al-muzlim', or camera obscura, is a sad omission to acknowledge from recent studies of Kitab al-Manazir, which should have been referred to by these writers (Smith, 2008). Further, such studies could have also mentioned Ibn al-Haytham's lack of realisation of such an obvious effect and why. One may, for example, say that he (Ibn al-Haytham) may have been concerned at this point in his research or was so focused on reporting on the properties of the travel of light rays that he did not want to mix the two research points together. However, if one looks at the complete body of work by Ibn al-Haytham, by reading this contribution in such a light, one may arrive at a different conclusion from what we have seen in recent years. My conclusion in this respect, as a practising scientist engaged in scientific instrument developments over several decades, is that many such writers lack the understanding of image formation. Further, it is also a lack of understanding on their behalf of how acknowledgements in science are accredited to all who contribute in the development of and/or understanding of a given celebrated discovery or an invention (see discussion above on the invention of the blue LED).

It is clear from the foregoing discussion that the work of Ibn al-Haytham in the field of optics, as an example, as well as of other Arab scholars, has enriched our world with their research work. In this context, their contributions should be viewed, if not celebrated, as evidence of the shared cultural roots of science. So to give credit where it is due, and as an example of this practice, should one associate Ibn al-Haytham with the invention and development of the camera obscura as an imaging instrument? I here wish Professor Roshdi Rashed a long and healthy life, and say in the tradition of the early Arab scholars whom we have discussed ‏– انساء الله في عمره-‎ (meaning may Allah add to his age) so he can finish his study of the burning spheres, so that others may revisit Kitab al-Manazir, the burning spheres and other works by this scholar to investigate al-Dil al Muzlim of Ibn al Haytham and his relationship to the imaging properties of the camera obscura.

ACKNOWLEDGEMENTS

The author would like to thank Professor Rashed for useful discussion, Professor Colin Humphrey for use of Figure 11.7, and Omar El-Gomati and Hatem Salih for a critical reading of the manuscript.

REFERENCES

Abū al-Ḥasan ‘Alī ibn Zayd al-Bayhaqī, 1976, *Tarikh Hukama' al-Islam*, ed. K Ali, Damascus, Arab League Academy of Damascus, second edition.

Al-Qifti, 1903, *Ta'rikh al-Hukama*, Leipzig, Lippert .

Dowed, J, 2012, *The Vitamin D Cure*, John Wiley and Sons, Hoboken, NJ.

El-Gomati, M M, Wells, T C R, Mullerova, I, Frank, L and Jakody, H, 2004, Why is it that differently doped regions in semiconductors are visible in low voltage SEM? *IEEE Trans. Electron Devices*; 51(2): 288–291.

El-Gomati, M M, Zaggout, F, Jayakody, H, Tear, S and Wilson, K, 2005, *Surf. Interface Anal.*; 37: 901–911.

Hubble telescope mission, 1990, http://www.nasa.gov/mission_pages/hubble/servicing/index.html

Ibn Abi Usaibi'a, 1965, *Uyun al-anba' fi tabaqaat al-atibba'*, ed. N Rida, Beirut, Institute for Arabic and Islamic Studies at the University of Frankfurt, Germany, pp. 90–98.

Krvanek, O J, Delby, N, Spence, A J Rodenberg, J M and Brown, L M, 1997, Aberration correction in the STEM, In *Proc. EMAG*, ed. J M Rodenberg, 35–39 (Institute of Physics Conf. Ser 153).

Nazif, M, 1942, *Al-Hasan ibn al-Haytham, buhuthuhu wa-kushufuhu al-basariyya*, al-Itimad Press, Cairo.

Public Health England Report, 2007, SACN Update on Vitamin D.

Rashed, R, 1990, A pioneer in anaclastics: Ibn Sahl on burning mirrors and lenses, *ISIS*; 81, 464–491.

Rashed, R, 1996, *Geomtire et Dioptrique au X' Siecle Ibn Sahl, Quhi et Ibn Al-Haytham*, Les Belles Lettres, Paris, pp. 36–40.

Rashed, R, 2005, *Geometry and Dioptrics in Classical Islam*, London, Alfurqan, pp. 1039–1043.

Rashed, R, 2013, *Ibn al-Haytham and Analytical Mathematic: A History of Arabic Sciences and Mathematics*, vol. 2, Routledge, London.

Sabra, A I ed., 1983, *The Optics of Ibn al-Haytham, Books I–II–III: On Direct Vision. The Arabic Text, Edited and with Introduction, Arabic-Latin Glossaries and Concordance Tables*, Kuwait, National Council for Culture, Arts and Letters.

Smith, A M, 2001, *Alhacen's Theory of Visual Perception: A Critical Edition, with English Translation and Commentary of the First Three Books of Alhacen's De Aspectibus, the Medieval Latin Version of Ibn al-Haytham's Kitab al-Manazir*, The American Philosophical Society, Philadelphia.

Smith, A M, ed. and trans. 2008, *Alhacen on Image-Formation and Distortion in Mirrors: A Critical Edition, with English Translation and Commentary, of Book 6 of Alhacen's De Aspectibus, the Medieval Latin Version of Ibn al-Haytham's Kitāb al-Manāzir*, Transactions of the American Philosophical Society, Philadelphia.

Spallas, J P, Silver, C S, Murray, L P, Wells, T and El-Gomati, M M, 2006, *Microelectronic Engineering*, Elsevier, Amsterdam, Netherlands, pp. 984–985.

The USA Department of Energy Report, 2014, Wind Technologies Market Report, U.S. Department of Commerce National Technical Information Service 5285 Port Royal Road Springfield, VA.

Zhihao, Y et al. 2014, Towards intrinsic charge transport in monolayer molybdenum disulfide by defect and interface engineering. *Nat. Commun.*; 5: 5290, doi: 10.1038/ncomms6290.

Zewail, A H, 2010, 4D Electron microscopy, *Science*; 328: 187.

Zewail A H, Thomas J M, 2009, *4D Electron Microscopy: Imaging in Space and Time*, Imperial College Press, London.

12 New Short-Wavelength Pulsed Light Sources

Majed Chergui

CONTENTS

Historical Introduction 165
Ultrafast Structural Dynamics with X-Rays: Sources 169
    Plasma-Based Sources 170
    Large-Scale Facilities: Synchrotrons 170
    Large-Scale Facilities: X-Ray Free Electron Lasers (XFEL) 172
References 174

HISTORICAL INTRODUCTION

The discoveries of Ibn al-Haytham a millennium ago, and in particular the establishment of the experimental methods based on the study of Optics,[1] have triggered an irresistible chain of discoveries over the following centuries, which would reach its culmination in the twentieth century and is still going on today.

It would require six centuries after Ibn-al-Haytham's fundamental works to demonstrate that visible light is composed of the colours of the rainbow, as was shown by Isaac Newton. However, it was later understood that light (or the electromagnetic spectrum) is not only composed of what is visible to the eye, but it also spans several decades of wavelengths (or frequencies). Indeed, about two centuries later, in 1800, infrared radiation was discovered by the astronomer (and also musician and composer of musical pieces) Sir Frederick William Herschel, born in Hanover (Germany), when he investigated the temperature of each colour of the solar spectrum passing through a prism. To do this, he used three blackened thermometers (to better absorb heat). He placed one thermometer directly in the colour part of the spectrum and the other two outside, as controls. He noticed that the temperature in the colour part was greater than that in the controls and that it increased from violet to red. However, he also noticed that the temperature of the region just beyond the red part of the spectrum was higher than the red itself! This was the discovery of infrared radiation. He also showed that it can be absorbed, transmitted, reflected and refracted.

A year later, in 1801, the pharmacist Johann Wilhelm Ritter, born in Samitz (now part of Poland), was experimenting with silver chloride. It was then known that exposure of this material to blue light caused a greater reaction than exposure to red light. He exposed silver chloride in each colour of the spectrum to determine the rate at which it reacted, but noticed that the silver chloride showed little change in the red part of the spectrum, whereas it became increasingly darker towards the violet end of the spectrum. He then placed silver chloride in the area just beyond the violet end of the spectrum, in a region where no light was visible, and surprisingly found that the silver chloride displayed an intense reaction. This marked the discovery of ultraviolet radiation. The experiments by Herschel and Ritter proved that invisible forms of radiation existed beyond both ends of the visible spectrum. The nineteenth century would then be marked by several discoveries in Optics, thanks to the works of Maxwell, Hertz, Fresnel, Young and many others. They would, among others, demonstrate the existence of electromagnetic radiation in other wavelength ranges, such as radio waves.

Extreme short wavelength radiation, i.e. x-rays, was discovered by William Conrad Röntgen in 1895. It sparked off a breath-taking chain of scientific break-throughs in the first two decades of the twentieth century that marked the birth of structural science. The systematic study of the characteristic rays (in emission and absorption) was mainly undertaken by Charles Glover Barkla, who discovered x-ray absorption edges (interestingly, this discovery preceded x-ray diffraction).[2,3] Soon after, the seminal works of Paul Ewald, Max von Laue, William Henry Bragg and William Lawrence Bragg in the period between 1912 and 1913 heralded the birth of a new Era by establishing the laws of x-ray diffraction from crystals and the determination of their structures with atomic resolution.[4-8] Soon afterwards, Debye investigated the effects of thermal motions and disorder showing that they decrease the diffraction intensities of Bragg reflections, especially at high scattering angles.[9] This would lead to the birth of the Debye–Scherrer scattering for powder diffraction and diffuse scattering, which applies to disordered media. Further additional discoveries, such as electron diffraction and scattering, made such that by the 1920s, one could resolved the structure of objects at the atomic scale of space, i.e. the Å. It can be said that while several spectacular improvements would come in the following decades, the stage was then set for the static structural determination of assemblies of atoms (crystals, molecules, proteins).

In parallel to the discoveries concerning the scattering of x-rays, work was pursued to understand and interpret x-ray absorption and emission spectra, as the development of quantum mechanics was unfolding. Manne Siegbahn[10-13] developed new apparata and methods to rigorously analyse x-ray spectra. However, x-ray spectroscopy would remain limited to chemical identification of elements and analysis of electronic structure. As a structural tool, x-ray absorption spectroscopy would have to go through several decades of theoretical developments (beautifully described in the historical review by Stumm von Bordwehr[14]) aimed at describing the modulations that appear in the above-edge region of the spectrum of an atom, when it is embedded in a molecular or crystalline edifice. These modulations, called the x-ray absorption near-edge structure (XANES) and extended x-ray absorption fine structure (EXAFS), finally received a consistent interpretation through the works of Sayers, Stern and Lytle.[15] Several improvements of the theoretical tools will ensue,[16] making EXAFS in particular (whose treatment is easier) an important element-selective tool in Materials Science, Biology and Chemistry.

In the investigation of the static nuclear structure of molecules, crystals and proteins, x-ray methods remain among the most popular tools given the huge instrumentation that is being built to this purpose. However, matter is not static and chemical reactions, phase transitions in solids or biological functions are the basis of life. The timescales of chemical and physical transformation in matter span several decades and the required time resolution depends on the scientific question one needs to address (Figure 12.1). Animal motion requires typically millisecond resolution, and the first such snapshots were taken by the French anatomist and physiologist, Etienne-Jules Marrey,[17] who developed the first shutter camera. Large amplitude protein motion requires ms to μs, which can be monitored by NMR or optical techniques. As the length scale of processes decreases, for example acoustic waves, molecular rotation, molecular vibrations, there is a need for higher and higher temporal resolution spanning the range from nanoseconds (ns) to femtoseconds (fs). For the description of processes occurring at the atomic scale of length, that is the chemical bond, fs-resolution is required. This is the timescale of nuclear motion in molecules, crystals and biosystems, and has, for that matter, been coined the atomic scale of time by A H Zewail, who pioneered Femtochemistry, by demonstrating the power of fs pump-probe spectroscopy to describe chemical processes. He was awarded the Nobel Prize for Chemistry in 1999 for his seminal contributions.[18-22]

Indeed, for the first time it became possible to probe in real time the motion of atoms inside molecules, via their wave-packet dynamics, and monitor chemical reactions and the transition state. This major development, and that of others in the growing community of ultrafast scientists,[23-28] was all carried out using optical pulses in the IR, visible and ultraviolet. Yet from the beginning, was it clear to many in this community that optical-domain radiation does not deliver structural information, except in a few rare cases, that is small molecules whose energy topology (i.e., potential energy curves) is known. This does not apply to assemblies of more than three atoms and therefore, early on in the 1990s, there were suggestions to replace the optical probe pulse with ultrashort pulses of electrons or of x-rays. A H Zewail was the sole to adopt the route with electron pulses in the early 1990s,[29-32] while most of the community adopted x-rays.[33-45] The main challenge in this endeavour was the development of sources of ultrashort electron or x-ray pulses and the mastering of data acquisition schemes which differed greatly depending on whether one carries out diffraction, scattering or spectroscopic measurements. The common technical needs of scientists coming from very different scientific backgrounds (Chemistry, Biology, Solid state physics, Materials Science, Engineering) has created a convergence of interests towards instrumentation that is used equally by different communities.

The pioneering work of the Zewail group on ultrafast electron diffraction and, more recently, microscopy and Electron Energy Loss Spectroscopy has been reviewed in several excellent papers and is beyond the scope of this review.[46-51] Rather, here we focus more specifically on the studies using x-ray pulses that have been undertaken in the past 15 years to probe the photoinduced structural changes in molecular systems, either in solutions or in crystalline form. While several reviews have already been published on the x-ray probing of molecular systems,[47,52-62] here we focus more on the methods.

Thus, probing the nuclear dynamics of assemblies of atoms requires both the atomic-scale resolution of space (the Å) of structural methods such as x-ray or electron diffraction and x-ray absorption spectroscopy, with the atomic resolution

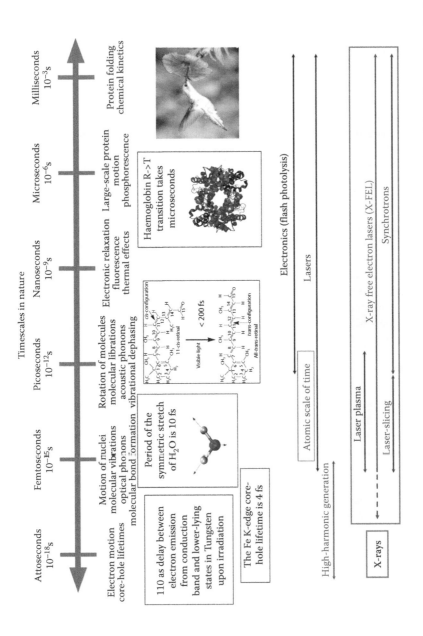

FIGURE 12.1 Timescales of fundamental processes in nature and (below) types of sources to probe them in real time. The atomic scale of time is the femtosecond and that of length is the Angström.

of time, the femtosecond. The principle of all these approaches consists in exciting the sample with an ultrashort laser pump pulse and probing it with a second ultrashort x-ray (or electron) probe pulse, whose time delay with respect to the first is tuned by optical delay lines in the femtosecond to picosecond range, or even longer times. Here, we will briefly review the most recent developments in the field of pulsed x-ray sources for the purpose of carrying out both static and time-resolved studies.

ULTRAFAST STRUCTURAL DYNAMICS WITH X-RAYS: SOURCES

Performing time-resolved x-ray experiments with picosecond to femtosecond time resolution requires ultrashort x-ray pulses, since there are no detectors capable of picosecond to femtosecond resolution. Thus, the time resolution has to rely on the source and on the use of optical delay lines, just as in ultrafast optical spectroscopy. The principle is based on the pump-probe method (Figure 12.2) where a first ultrashort (pump) optical pulse triggers a physical, chemical or biochemical process in the system of interest, while a second ultrashort x-ray (probe) pulse will probe the system at variable time delays after the pump pulse. In the x-ray domain, the measured signal can be x-ray absorption, x-ray emission, x-ray scattering or x-ray diffraction, depending on the type of sample (solid or solutions) and on the properties (electronic, structural) one wants to investigate.

There are several classes of sources of ultrashort x-ray pulses, which can be broadly classified as tabletop or large-scale installations. Within these two categories, sources also differ in pulse duration, energy range, energy tunability, etc. Among the tabletop sources that are commonly being used, one counts: (i) laser-driven tabletop plasma sources that can generate 100 fs hard x-ray pulses[63]; (ii) high harmonic generation (HHG) sources that can reach attosecond pulse durations but are limited to the vacuum ultraviolet (VUV) to soft x-ray range.[64] The large-scale installations are all accelerator-based facilities such as synchrotrons, which can generate 100 ps x-ray pulses that can be shortened to 100 fs using the slicing scheme,[65–68] or x-ray free electron lasers (X-FELs), which can generate intense x-ray pulses of few fs.[69] In the following paragraphs, we will give a brief description of ultrashort pulsed hard x-ray

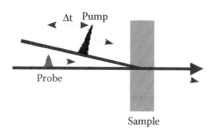

FIGURE 12.2 Principle of the pump-probe method. The pump pulse triggers a process in the sample, which is then monitored by a probe pulse whose time delay is tunable with respect to the pump pulse.

sources based on tabletop and large-scale installations, as several detailed reviews on each type can already be found in the literature.[54,58,63,69–71] The most commonly used sources to date have been the plasma-based ones (tabletop), synchrotrons and x-ray free electron laser, on which we concentrate hereafter.

PLASMA-BASED SOURCES

In laser-driven plasma sources,[72] a femtosecond laser pulse with peak intensity higher than 10^{16} W/cm^2 is used to irradiate a metal target, resulting in generation of a plasma. The acceleration of free electrons into the target by the very high electric field of the pulses and the generation of characteristic radiation and Bremsstrahlung by interactions with target atoms emits radiation in the soft and hard x-ray ranges. However, the emitted x-ray pulse is isotropic and it fluctuates in intensity on a pulse-to-pulse basis, requiring special optics to collect the radiation and special set-up designs to correct for the fluctuations. Usually, the collected radiation amounts to only a small part of the typically 10^9 x-ray photons emitted per pulse, because of constraints on time resolution.[63] For a target thickness of the order of 10 μm, the characteristic emission has a duration of the order of 100 fs. An interesting intrinsic property of this technique is that time zero is very well defined, as the same optical laser is used to excite the sample and generate the plasma.

Recent developments in laser and target technologies have allowed the generation of hard x-ray pulses with kHz repetition rates and much smaller intensity fluctuations. Since the irradiated target is damaged irreversibly, moving targets providing a fresh area for each laser pulse are required. In addition, a highly stable spatial position of the target area with fluctuations of less than 20 μm is needed. Some hard x-ray plasma sources with Cu tape targets are driven by sub-50 fs pulses and work at a 1 kHz repetition rate[73] and the collimated hard x-ray flux on the sample reaches up to 10^7 photons/s. It was shown that by using intense mid-infrared sub-100-fs pulses (3.9 μm), the comparably long optical period allows for accelerating electrons from the Cu target to much higher kinetic energies, thus generating an x-ray flux 25 times higher than with 800 nm laser pulses.[74]

LARGE-SCALE FACILITIES: SYNCHROTRONS

The x-ray pulses generated by third-generation synchrotrons are of the order of 50–150 picosecond, and can be used to probe photoinduced structural changes on timescales of >100 ps by x-ray diffraction, scattering or spectroscopy. Synchrotrons (Figure 12.3) are circular accelerators where the charged particles (electrons or positrons) are accelerated to relativistic speeds. Upon deflection of the beam at a bending magnet or at insertion devices, radiation is emitted which spans the spectral range from the infrared to the x-rays, depending on the parameters of the machine.

The pioneering development of the x-ray scattering technique with 100 ps time resolution was performed at the European Synchrotron Radiation Facility (ESRF, France).[75,76] It is now available in several places around the world, such as the APS (Argonne, USA) and the KEK-AR (Tsukuba, Japan).[77]

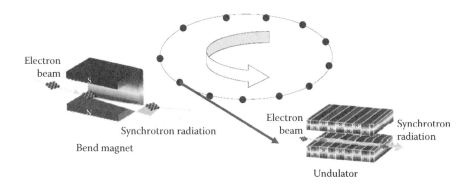

FIGURE 12.3 Operating principle of a synchrotron. Electron (or positron) bunches are stored in a circularly shaped storage ring and are accelerated to relativistic speeds. Every time the electron bunch is deflected, it emits electromagnetic radiation. There are two types of devices to deflect the bunches: classically, deflection magnets were used, but now insertion devices such as wigglers or undulators are used, which make the electron bunch wiggle in an array of north-south magnets, leading to a build-up of the flux and a higher coherence. The size of the electron bunches determines the time duration of the emitted pulse of light, which is 100 ps typically.

X-ray absorption spectroscopy with a 50–100 ps resolution was pioneered by the Chergui group.[45,47,52,53,58,59,62,78–86] Parallel to the latter, developments driven by L Chen and her group were also taking place.[55,56,87–89] Time-resolved x-ray spectroscopy at synchrotrons quickly emerged as an important tool for studying solution phase photoinduced electronic and structural changes in chemical systems in Germany, Italy, the United States,[90,91] Japan,[77,92,93] and Switzerland.

Until about 2010, these experiments were carried out using a 1 kHz femtosecond laser, while synchrotrons operate at MHz repetition rates, so that at least 10^3 of the x-ray pulses remained unused. A major development was introduced by the Chergui group, which consisted in using a ps pump laser (rather than an fs one) whose repetition rate is half or an integer fraction of that of the synchrotron.[94] Keeping all other parameters fixed, this leads to an increase of the signal by the root square of the laser repetition rate compared to the same signal recorded with a 1 kHz laser. This scheme also opens the possibility to perform photon-in/photon-out experiments (including scattering and diffraction), which require a high incident flux per second.[95] This development was followed by similar ones at the APS[96] and Elettra.[97]

In order to extract femtosecond x-ray pulses from synchrotrons, in particular with energy tuneability, the slicing scheme was proposed by Zholents and Zholterev[98] and its experimental validation was reported at the ALS.[65] The slicing technique (Figure 12.4) uses the very high electric field achievable by the fs laser pulse to modulate the energy of a 'slice' (100 fs) of an electron bunch (100 ps) in the storage ring. This slice has a larger energy spread than the rest of the electrons in the bunch, and thereby its emission deviates when passing through a radiator. This is because the electrons entering the insertion device with different energies will follow different paths. The beamlines first developed at the ALS[99,100] and at BESSY Germany[101] were operating in the soft x-ray range. The first hard x-ray slicing source was implemented

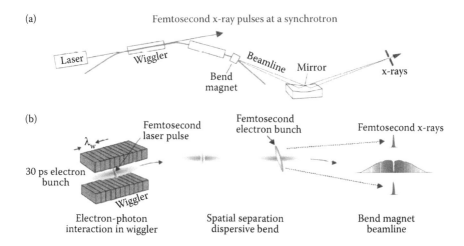

FIGURE 12.4 Schematic illustration of the method for generating femtosecond x-rays from a storage ring. (a) A femtosecond laser pulse energy modulates the electron bunch as it co-propagates through a resonantly tuned wiggler. (b) The dispersive section of the storage ring results in a spatial separation of the energy-modulated electron slice. (c) X-rays generated by the modulated electron bunch (in a bend magnet) are collected and imaged onto a slit that is used to select the femtosecond x-rays (generated off-axis) or the femtosecond dark pulse (generated on-axis). (Reproduced from Andrea Cavalleri A; Schoenlein, R W: *Topics in Applied Physics* 2004, 92, 309–338.)

at the Swiss Light Source,[68] where a fs laser pulse is split to 'pump' and 'probe' branches. The 'probe' branch modulates the energy of the electrons in the modulator where electrons in the magnetic device interact with the femtosecond laser. The slice of electrons passing through the undulator generates a femtosecond x-ray pulse used to probe the sample at a variable time delay with respect to the pump pulse which is intrinsically synchronised to the sliced beam. This is a great advantage of the slicing scheme, but the photon flux in the x-ray pulse is very low (typically 1/1,000 of the 50–100 ps pulse), even more so when performing x-ray absorption spectroscopy experiments, which require a narrow energy bandpass. Nevertheless, it was with the slicing scheme that the first fs x-ray spectroscopy experiment on a molecular system was achieved.[86] The new generation of machines, x-ray free electron lasers, overcomes this limitation as discussed hereafter.

LARGE-SCALE FACILITIES: X-RAY FREE ELECTRON LASERS (XFEL)

X-ray Free Electron Lasers are linear accelerators with insertion devices (undulators) at their end, so that the electron bunches are used only once (Figure 12.5). The magnetic structure, used to generate x-rays, consist of linear sections hundreds of metres long. Such long interaction volumes force the electrons in the bunch to interact constructively via the generated x-ray field, giving rise to the so-called micro-bunching, which shortens the x-ray pulse duration tremendously ($\leq$100 fs), enhances coherence, emittance and, importantly, flux, with values of 10^{12} photons/pulse in

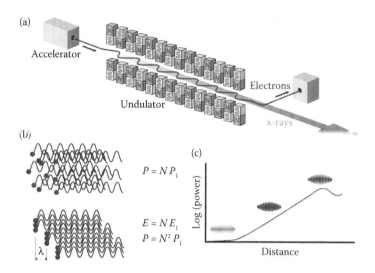

FIGURE 12.5 Operation principle of the Free Electron Laser. Electrons are accelerated close to the speed of light. As the beam passes through an undulator (a), which is an array of magnets, the electrons are forced to wiggle transversely along a sinusoidal path about the axis of the undulator, leading to emission of light. When the latter becomes sufficiently strong, interaction between the transverse electric field of the light and the transverse electron current causes some electrons to gain and others to lose energy to the optical field. This energy modulation leads to a microbunching (b) of the electrons, which are separated by one optical wavelength along the axis, and to an increase in radiated power and coherence (c).

0.1% bandwidth. This self-amplified spontaneous emission (SASE) of hard x-rays was generated for the first time in the hard x-ray range at the Linac Coherent Light Source (LCLS) at SLAC (Stanford, USA)[102] and later at SACLA (Sayo, Japan).[103] New machines will be inaugurated in 2017: the European X-FEL (Hamburg, Germany) and the SwissFEL at the Paul Scherrer Institute (Villigen, Switzerland), and PAL (Korea).

The LCLS and SACLA deliver up to 10^{14} photons/second on the sample and the time duration of the x-ray pulses can be decreased down to a few fs. Owing to the fluctuating initial conditions of the SASE, the spectral and the time structures of the x-ray pulses fluctuate on a shot-to-shot basis. However, it is now possible to correct for these fluctuations using monochromators and normalisation methods for better characterising the x-ray spectrum, while the use of the timing tool allows achieving few-fs time sorting.[104]

While the hard x-ray FELs are all based on the principle of SASE, one machine, FERMI@Elettra (Trieste, Italy), operates as a seeded FEL and therefore as a real laser.[105–108] It generates soft x-ray pulses up to about 300 eV, and work is in progress to increase this limit to the oxygen K-edge.

In summary, the past 10 years have witnessed a huge development in pulsed x-ray sources. Here, we focused mainly on the hard x-ray sources, but several developments are still ongoing at large-scale installations and in labs to develop soft x-ray FELs or tabletop sources.

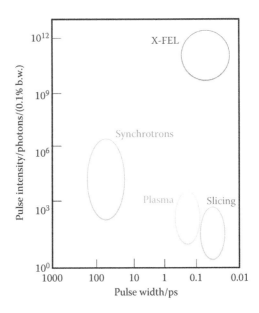

FIGURE 12.6   Comparison of fluxes of the different sources of ultrashort hard (2–20 keV) x-ray pulses, together with their temporal pulse widths.

Figure 12.6 summarises the present status of pulsed hard x-ray sources in terms of flux and time duration. Needless to say, the development of XFELs has been a game changer and the construction of several such machines around the world is heralding a new Era of structural dynamics studies.

REFERENCES

1. Rashed, R Portraits of science: A polymath in the 10th century. *Science* 2002, 297, 773–773.
2. Barkla, C G; Sadler, C A: The absorption of X rays. *Nature* 1909, 80, 37–37.
3. Barkla, C G; Collier, V: The absorption of X-rays and fluorescent X-ray spectra. *Philosophical Mag* 1912, 23, 987–997.
4. Friedrich, W; Knipping, P; Laue, M: Interference appearances in X-rays. *Ann Phys-Berlin* 1913, 41, 971–988.
5. Friedrich, W: A new interference occurrence with x-radiation. *Physikalische Zeitschrift* 1913, 14, 317–319.
6. Bragg, W H: X rays and crystals. *Nature* 1913, 90, 219–219.
7. Bragg, W H; Bragg, W L: The reflexion of x-rays through crystals. *Z Anorg Chem* 1914, 90, 169–181.
8. Bragg, W H: The reflexion of x-rays through crystals (II). *Z Anorg Chem* 1914, 90, 182–184.
9. Debye, P: Interference of x rays and heat movement. *Ann Phys-Berlin* 1913, 43, 49–95.
10. Siegbahn, M: On the spectrums of high frequency. *Comptes Rendus Hebdomadaires Des Seances De L Academie Des Sciences* 1917, 165, 59–59.
11. Siegbahn, M: Precision-measurements in the X-ray spectra. Part II. *Philosophical Mag* 1919, 38, 639–646.

12. Siegbahn, M; Jonsson, E: The absorption cut-off frequency of the x-rays among the heavy elements, especially among the rare earths. *Physikalische Zeitschrift* 1919, 20, 251–256.

13. Siegbahn, M; Lindh, A E; Stensson, N: On a method of spectral analysis using x-rays. *Zeitschrift Fur Physik* 1921, 4, 61–67.

14. Stumm von Bordwehr, R: A History of X-ray absorption fine structure. *Ann Phys Fr* 1989, 14, 377–465.

15. Sayers, D E; Stern, E A; Lytle, F W: New Technique for Investigating Noncrystalline Structures – Fourier Analysis of Extended X-Ray – Absorption Fine Structure. *Phys Rev Lett* 1971, 27, 1204.

16. Koningsberger, D C; Prins, R: *X-Ray Absorption: Principles, Applications, Techniques of EXAFS, SEXAFS, and XANES*; Wiley: New York, 1988.

17. Marrey, E-J: Du vol des oiseaux. *La Revue scientifique* 14, 21 August, 11 September and 2 October, 1869.

18. Zewail, A H: Femtosecond transition-state dynamics. *Faraday Discuss* 1991, 91, 207–237.

19. Zewail, A, ed.: *The Chemical Bond Structure and Dynamics*; Academic press: Boston, San Diego, New York, 1992.

20. Dantus, M; Zewail, A: Introduction: Femtochemistry. *Chem Rev* 2004, 104, 1717–1718.

21. Zewail, A H: Femtochemistry: Atomic-scale dynamics of the chemical bond using ultrafast lasers – (Nobel lecture). *Angew Chem Int Edit* 2000, 39, 2587–2631.

22. Zewail, A H: Femtochemistry: Atomic-scale dynamics of the chemical bond. *J Phys Chem A* 2000, 104, 5660–5694.

23. Fleming, G R; Cho, M H: Chromophore-solvent dynamics. *Annu Rev Phys Chem* 1996, 47, 109–134.

24. Jung, G; Ma, Y Z; Prall, B S; Fleming, G R: Ultrafast fluorescence depolarisation in the yellow fluorescent protein due to its dimerisation. *Chemphyschem* 2005, 6, 1628–1632.

25. Pugliano, N; Gnanakaran, S; Hochstrasser, R M: The dynamics of photodissociation reactions in solution. *J Photoch Photobio A* 1996, 102, 21–28.

26. Zewail, A: *Femtochemistry Ultrafast Dynamics of the Chemical Bond*; World scientific: Singapore, New Jersey, London, 1994.

27. Chergui, M: Femtochemistry: Ultrafast chemical and physical processes in molecular systems, *Femtochemistry: The Lausanne Conference*, Lausanne, Switzerland, September 4–8, 1995; World Scientific: Singapore; River Edge, NJ, 1996.

28. Sundström, V: Femtochemistry and femtobiology: ultrafast reaction dynamics at atomic-scale resolution: Nobel Symposium 101; Imperial College Press; Distributed by World Scientific Pub. Co.: London. River Edge, NJ, 1997.

29. Williamson, J C; Zewail, A H: Structural femtochemistry – experimental methodology. *P Natl Acad Sci USA* 1991, 88, 5021–5025.

30. Williamson, J C; Dantus, M; Kim, S B; Zewail, A H: Ultrafast diffraction and molecular-structure. *Chem Phys Lett* 1992, 196, 529–534.

31. Williamson, J C; Zewail, A H: Ultrafast electron-diffraction .4. Molecular-structures and coherent dynamics. *J Phys Chem* 1994, 98, 2766–2781.

32. Dantus, M; Kim, S D; Williamson, J C; Zewail, A H: Ultrafast electron diffraction .5. Experimental time resolution and applications. *J Phys Chem* 1994, 98, 2782–2796.

33. Ma, H; Lin, S H; Rentzepis, P: Theoretical-study of laser-heating and dissociation reactions in solids using ultrafast time-resolved X-ray-diffraction. *J Appl Phys* 1992, 72, 2174–2178.

34. Anderson, T; Tomov, I V; Rentzepis, P M: A high-repetition-rate, picosecond hard x-ray system, and its application to time-resolved x-ray-diffraction. *J Chem Phys* 1993, 99, 869–875.

35. Kieffer, J C; Chaker, M; Matte, J P; Pepin, H; Cote, C Y; Beaudoin, Y; Johnston, T W et al.: Ultrafast x-ray sources. *Phys Fluids B-Plasma* 1993, 5, 2676–2681.

36. Petkov, V; Takeda, S; Waseda, Y; Sugiyama, K: Structural study of molten germanium by energy-dispersive x-ray-diffraction. *J Non-Cryst Solids* 1994, 168, 97–105.

37. Tomov, I V; Chen, P; Rentzepis, P M: Nanosecond hard x-ray source for time-resolved x-ray-diffraction studies. *Rev Sci Instrum* 1995, 66, 5214–5217.

38. Chen, P; Tomov, I V; Rentzepis, P M: Time resolved heat propagation in a gold crystal by means of picosecond x-ray diffraction. *J Chem Phys* 1996, 104, 10001–10007.

39. Larsson, J; Judd, E; Falcone, R W; Asfaw, A; Lee, R W; Heimann, P A; Padmore, H A; Wark, J: Time-resolved x-ray diffraction from laser-heated crystals. *Inst Phys Conf Ser* 1996, 151, 367–371.

40. Wulff, M; Ursby, T; Bourgeois, D; Schotte, F; Zontone, F; Lorenzen, M: New opportunities for time resolved x-ray scattering at the ESRF. *J Chim Phys Pcb* 1996, 93, 1915–1937.

41. Larsson, J; Chang, Z; Judd, E; Schuck, P J; Falcone, R W; Heimann, P A; Padmore, H A et al.: Ultrafast x-ray diffraction using a streak-camera detector in averaging mode. *Opt Lett* 1997, 22, 1012–1014.

42. Raksi, F; Wilson, K R; Jiang, Z M; Ikhlef, A; Cote, C Y; Kieffer, J C: Ultrafast x-ray absorption probing of a chemical reaction. *J Chem Phys* 1996, 104, 6066–6069.

43. BenNun, M; Cao, J S; Wilson, K R: Ultrafast x-ray and electron diffraction: Theoretical considerations. *J Phys Chem A* 1997, 101, 8743–8761.

44. Cao, J S; Wilson, K R: Ultrafast x-ray diffraction theory. *J Phys Chem A* 1998, 102, 9523–9530.

45. Bressler, C; Chergui, M; Pattison, P; Wulff, M; Filipponi, A; Abela, R: A laser and synchrotron radiation pump-probe x-ray absorption experiment with sub-ns resolution. *Proc SPIE* 1998, 3451, 108–116.

46. Shorokhov, D; Zewail, A H: 4D electron imaging: Principles and perspectives. *Phys Chem Chem Phys* 2008, 10, 2879–2893.

47. Chergui, M; Zewail, A H: Electron and x-ray methods of ultrafast structural dynamics: Advances and applications. *Chemphyschem* 2009, 10, 28–43.

48. Zewail, A; Thomas, J M: *4D Electron Microscopy Imaging in Space and Time*; Imperial college press: London, 2010.

49. Yang, D S; Baum, P; Zewail, A H: Ultrafast electron crystallography of the cooperative reaction path in vanadium dioxide. *Struct Dyn* 2016, 3, 034304.

50. van der Veen, R M; Penfold, T J; Zewail, A H: Ultrafast core-loss spectroscopy in four-dimensional electron microscopy. *Struct Dyn* 2015, 2, 024302.

51. Shorokhov, D; Zewail, A H: Perspective: 4D ultrafast electron microscopy-evolutions and revolutions. *J Chem Phys* 2016, 144.

52. Bressler, C; Chergui, M: Ultrafast X-ray absorption spectroscopy. *Chem Rev* 2004, 104, 1781–1812.

53. Bressler, C; Abela, R; Chergui, M: Exploiting EXAFS and XANES for time-resolved molecular structures in liquids. *Z Kristallogr* 2008, 223, 307–321.

54. Bressler, C; Chergui, M: Time-resolved X-ray absorption spectroscopy. *Actual Chim* 2008, 317, 59–61.

55. Chen, L X: Probing transient molecular structures in photochemical processes using laser-initiated time-resolved x-ray absorption spectroscopy. *Annu Rev Phys Chem* 2005, 56, 221–254.

56. Chen, L X; Zhang, X Y; Lockard, J V; Stickrath, A B; Attenkofer, K; Jennings, G; Liu, D J: Excited-state molecular structures captured by x-ray transient absorption spectroscopy: A decade and beyond. *Acta Crystallogr A* 2010, 66, 240–251.

57. Graber, T; Anderson, S; Brewer, H; Chen, Y S; Cho, H S; Dashdorj, N; Henning, R W et al.: BioCARS: A synchrotron resource for time-resolved x-ray science. *J Synchrotron Radiat* 2011, 18, 658–670.

58. Bressler, C; Chergui, M: Molecular structural dynamics probed by ultrafast x-ray absorption spectroscopy. *Annual Rev Phys Chem* 2010, 61, 263–282.

59. Chergui, M: Picosecond and femtosecond x-ray absorption spectroscopy of molecular systems. *Acta Crystallogr A* 2010, 66, 229–239.

60. Chergui, M: In-situ characterization of molecular processes in liquids by ultrafast x-ray absorption spectroscopy. In *In-situ Materials Characterization*; Ziegler, A, Graafsma, H, Zhang, X F, Frenken, J W M, Eds.; Springer: Berlin, Heidelberg, 2014; Vol. 193; pp 1–38.

61. Milne, C J; Penfold, T J; Chergui, M: Recent experimental and theoretical developments in time-resolved x-ray spectroscopies. *Coordin Chem Rev* 2014, 277–278, 44–68.

62. Chergui, M: Time-resolved X-ray spectroscopies of chemical systems: New perspectives. *Struct Dyn* 2016, 3, 031001.

63. Elsaesser, T; Woerner, M: Photoinduced structural dynamics of polar solids studied by femtosecond x-ray diffraction. *Acta Crystallogr A* 2010, 66, 168–178.

64. Pfeifer, T; Spielmann, C; Gerber, G: Femtosecond x-ray science. *Rep Prog Phys* 2006, 69, 443–505.

65. Schoenlein, R W; Chattopadhyay, S; Chong, H H W; Glover, T E; Heimann, P A; Shank, C V; Zholents, A A; Zolotorev, M S: Generation of femtosecond pulses of synchrotron radiation. *Science* 2000, 287, 2237–2240.

66. Heimann, P A; Padmore, H A; Schoenlein, R W: ALS Beamline 6.0 for ultrafast x-ray absorption spectroscopy. *Synchrotron Radiat Instrum* 2004, 705, 1407–1410.

67. Khan, S; Holldack, K; Kachel, T; Mitzner, R; Quast, T: Femtosecond undulator radiation from sliced electron bunches. *Phys Rev Lett* 2006, 97, 074801.

68. Beaud, P; Johnson, S L; Streun, A; Abela, R; Abramsohn, D; Grolimund, D; Krasniqi, F; Schmidt, T; Schlott, V; Ingold, G: Spatiotemporal stability of a femtosecond hard-X-ray undulator source studied by control of coherent optical phonons. *Phys Rev Lett* 2007, 99 , 174801.

69. Bostedt, C; Boutet, S; Fritz, D M; Huang, Z; Lee, H J; Lemke, H T; Robert, A; Schlotter, W F; Turner, J J; Williams, G J: Linac coherent light source: The first five years. *Rev Mod Phys* 2016, 88, 015007.

70. Rousse, A; Rischel, C; Gauthier, J C: Ultrafast x-ray sources and applications. *Comptes Rendus De L Academie Des Sciences Serie Iv Physique Astrophysique* 2000, 1, 305–315.

71. Von der Linde, D; Sokolowski-Tinten, K; Blome, C; Dietrich, C; Zhou, P; Tarasevitch, A; Cavalleri, A et al.: Generation and application of ultrashort x-ray pulses. *Laser Part Beams* 2001, 19, 15–22.

72. Rousse, A; Rischel, C; Fourmaux, S; Uschmann, I; Sebban, S; Grillon, G; Balcou, P et al.: Non-thermal melting in semiconductors measured at femtosecond resolution. *Nature* 2001, 410, 65–68.

73. Zhavoronkov, N; Schmising, K V K; Bargheer, M; Woerner, M; Elsaesser, T; Klimo, O; Limpouch, J: High repetition rate ultrafast X-ray source from the fs-laser-produced-plasma. *J De Physique Iv* 2006, 133, 1201–1203.

74. Weisshaupt, J; Juve, V; Holtz, M; Ku, S A; Woerner, M; Elsaesser, T; Alisuskas, S; Pugzlys, A; Baltuska, A: High brightness table top hard X ray source driven by sub 100-femtosecond mid-infrared pulses. *Nat Photonics* 2014, 8, 927–930.

75. Wulff, M; Kong, Q Y; Lee, J H; Lo Russo, M; Kim, T K; Lorenc, M; Cammarata, M et al.: Photolysis of Br(2) in CCl(4) studied by time-resolved x-ray scattering. *Acta Crystallogr A* 2010, 66, 252–260.

76. Cammarata, M; Levantino, M; Schotte, F; Anfinrud, P A; Ewald, F; Choi, J; Cupane, A; Wulff, M; Ihee, H: Tracking the structural dynamics of proteins in solution using time-resolved wide-angle x-ray scattering. *Nat Methods* 2008, 5, 881–886.

77. Nozawa, S; Adachi, S I; Takahashi, J I; Tazaki, R; Guerin, L; Daimon, M; Tomita, A et al.: Developing 100 ps-resolved x-ray structural analysis capabilities on beamline NW14A at the photon factory advanced ring. *J Synchrotron Radiat* 2007, 14, 313–319.

78. Bressler, C; Saes, M; Chergui, M; Abela, R; Pattison, P: Optimizing a time-resolved X-ray absorption experiment. *Nucl Instrum Methods Phys Res Sect a-Accel Spectrom Detect Assoc Equip* 2001, 467, 1444–1446.

79. Bressler, C; Saes, M; Chergui, M; Grolimund, D; Abela, R; Pattison, P: Towards structural dynamics in condensed chemical systems exploiting ultrafast time-resolved x-ray absorption spectroscopy. *J Chem Phys* 2002, 116, 2955–2966.

80. Saes, M; Bressler, C; Abela, R; Grolimund, D; Johnson, S L; Heimann, P A; Chergui, M: Observing photochemical transients by ultrafast x-ray absorption spectroscopy. *Phys Rev Lett* 2003, 90, 047403–047411.

81. Saes, M G W; Kaiser, M; Tarnovsky, A; Bressler, Ch; Chergui, M; Johnson, S L; Grolimund, D; Abela, R: Ultrafast time-resolved x-ray absorption spectroscopy of chemical systems. *Synchrotron Radiat News* 2003, 16, 12.

82. Chergui, M; Bressler, C; Abela, R: The Fast show with X-rays and Electrons. *Synchrotron Radiat News* 2004, 17, 11–14.

83. Saes, M; van Mourik, F; Gawelda, W; Kaiser, M; Chergui, M; Bressler, C et al.: A setup for ultrafast time-resolved x-ray absorption spectroscopy. *Rev Sci Instrum* 2004, 75, 24–30.

84. Gawelda, W; Bressler, C; Saes, M; Kaiser, M; Tarnovsky, A N; Grolimund, D; Johnson, S L; Abela, R; Chergui, M: Picosecond time-resolved x-ray absorption spectroscopy of solvated organometallic complexes. *Physica Scripta* 2005, T115, 102–106.

85. Gawelda, W; Pham, V T; Benfatto, M; Zaushitsyn, Y; Kaiser, M; Grolimund, D; Johnson, S L et al.: Structural determination of a short-lived excited iron(II) complex by picosecond x-ray absorption spectroscopy. *Phys Rev Lett* 2007, 98, 057401.

86. Cannizzo, A; Milne, C J; Consani, C; Gawelda, W; Bressler, C; van Mourik, F; Chergui, M: Light-induced spin crossover in Fe(II)-based complexes: The full photocycle unraveled by ultrafast optical and x-ray spectroscopies. *Coordin Chem Rev* 2010, 254, 2677–2686.

87. Chen, L X: Probing transient molecular structures with time-resolved pump/probe XAFS using synchrotron X-ray sources. *J Electron Spectrosc Relat Phenom* 2001, 119, 161–174.

88. Chen, L X; Jager, W J H; Jennings, G; Gosztola, D J; Munkholm, A; Hessler, J P: Capturing a photoexcited molecular structure through time-domain x-ray absorption fine structure. *Science* 2001, 292, 262–264.

89. Jennings, G; Jager, W J H; Chen, L X: Application of a multi-element Ge detector in laser pump/x-ray probe time-domain x-ray absorption fine structure. *Rev Sci Instrum* 2002, 73, 362–368.

90. Huse, N; Kim, T K; Jamula, L; McCusker, J K; de Groot, F M F; Schoenlein, R W: Photo-induced spin-state conversion in solvated transition metal complexes probed via time-resolved soft x-ray spectroscopy. *J Am Chem Soc* 2010, 132, 6809–6816.

91. Katz, J E; Zhang, X Y; Attenkofer, K; Chapman, K W; Frandsen, C; Zarzycki, P; Rosso, K M; Falcone, R W; Waychunas, G A; Gilbert, B: Electron small polarons and their mobility in iron (Oxyhydr)oxide nanoparticles. *Science* 2012, 337, 1200–1203.

92. Sato, T; Nozawa, S; Ichiyanagi, K; Tomita, A; Chollet, M; Ichikawa, H; Fujii, H; Adachi, S-i; Koshihara, S-y: Capturing molecular structural dynamics by 100 ps time-resolved x-ray absorption spectroscopy. *J Synchrotron Radiat* 2009, 16, 110–115.

93. Nozawa, S; Sato, T; Chollet, M; Ichiyanagi, K; Tomita, A; Fujii, H; Adachi, S; Koshihara, S: Direct probing of spin state dynamics coupled with electronic and structural modifications by picosecond time-resolved XAFS. *J Am Chem Soc* 2010, 132, 61–63.

94. Lima, F A; Milne, C J; Amarasinghe, D C V; Rittmann-Frank, M H; van der Veen, R M; Reinhard, M; Pham, V T et al.: A high-repetition rate scheme for synchrotron-based picosecond laser pump/x-ray probe experiments on chemical and biological systems in solution. *Rev Sci Instrum* 2011, 82, 063111.

95. Bressler, C; Gawelda, W; Galler, A; Nielsen, M M; Sundstrom, V; Doumy, G; March, A M; Southworth, S H; Young, L; Vanko, G: Solvation dynamics monitored by combined x-ray spectroscopies and scattering: Photoinduced spin transition in aqueous [Fe(bpy)3]2+. *Faraday Discuss* 2014, 171, 169–178.

96. March, A M; Stickrath, A, Doumy, G; Kanter, E P; Krassig, B; Southworth, S H; Attenkofer, K; Kurtz, C A; Chen, L X; Young, L: Development of high-repetition-rate laser pump/x-ray probe methodologies for synchrotron facilities. *Review of Scientific Instruments* 2011, 82, 073110.

97. Stebel, L; Malvestuto, M; Capogrosso, V; Sigalotti, P; Ressel, B; Bondino, F; Magnano, E; Cautero, G; Parmigiani, F: Time-resolved soft x-ray absorption setup using multibunch operation modes at synchrotrons. *Rev Sci Instrum* 2011, 82, 123109.

98. Zholents, A A; Zolotorev, M S: Femtosecond x-ray pulses of synchrotron radiation. *Phy Rev Lett* 1996, 76, 912–915.

99. Steier, C; Robin, D; Sannibale, F; Schoenlein, R; Wan, W; Wittmer, W; Zholents, A: The new undulator based fs-slicing beamline at the ALS. *Ieee Part Acc Conf* 2005, 1645–1647.

100. Heimann, P A; Glover, T E; Plate, D; Lee, H J; Brown, V C; Padmore, H A; Schoenlein, R W: The advanced light source (ALS) slicing undulator beamline. *Aip Conf Proc* 2007, 879, 1195–1197.

101. Holldack, K; Khan, S; Mitzner, R; Quast, T: Femtosecond terahertz radiation from femtoslicing at BESSY. *Phys Rev Lett* 2006, 96, 054801.

102. Emma, P; Akre, R; Arthur, J; Bionta, R; Bostedt, C; Bozek, J; Brachmann, A et al.: First lasing and operation of an angstrom-wavelength free-electron laser. *Nat Photonics* 2010, 4, 641–647.

103. Ishikawa, T; Aoyagi, H; Asaka, T; Asano, Y; Azumi, N; Bizen, T; Ego, H et al.: A compact x-ray free-electron laser emitting in the sub-angstrom region. *Nat Photonics* 2012, 6, 540–544.

104. Harmand, M; Coffee, R; Bionta, M R; Chollet, M; French, D; Zhu, D; Fritz, D M et al.: Achieving few-femtosecond time-sorting at hard x-ray free-electron lasers. *Nat Photonics* 2013, 7, 215–218.

105. Allaria, E; Callegari, C; Cocco, D; Fawley, W M; Kiskinova, M; Masciovecchio, C; Parmigiani, F: The FERMI@Elettra free-electron-laser source for coherent x-ray physics: Photon properties, beam transport system and applications. *New J Phys* 2010, 12, 075002.

106. Zangrando, M; Cudin, I; Fava, C; Gerusina, S; Gobessi, R; Godnig, R; Rumiz, L; Svetina, C; Parmigiani, F; Cocco, D: First results from the commissioning of the FERMI@Elettra free electron laser by means of the Photon Analysis Delivery and Reduction System (PADReS). *Adv X-Ray Free Electron Lasers: Radiat Schemes, X-Ray Opt Instrument* 2011, 8078.

107. Allaria, E; Appio, R; Badano, L; Barletta, W A; Bassanese, S; Biedron, S G; Borga, A et al.: Highly coherent and stable pulses from the FERMI seeded free-electron laser in the extreme ultraviolet. *Nat Photonics* 2012, 6, 699–704.

108. Hara, T: Fully coherent soft x-rays at FERMI. *Nat Photonics* 2013, 7, 851–854.

109. Andrea Cavalleri, A; Schoenlein, R W: Femtosecond X-rays and structural dynamics in condensed matter. *Topics in Applied Physics* 2004, 92, 309–338.

13 Lighting
From Human Evolution to Sustainable Revolution

Harry Verhaar

CONTENTS

Introduction 181
Evolution: The Early Impact of Lighting 181
Historic Milestones 182
The First Revolution: Electric Light in the Home and Workplace 182
Challenges at Hand 184
Second Revolution: The Promise of LED 186
Connected Cities 187
Connected Offices 188
Connected Homes 189
Indoor Positioning 190
Eliminating Light Poverty 190
Something More than Illumination 191

INTRODUCTION

Some of the most important things in life are taken for granted, and nowhere is this truer than with light. Since man first beheld the rising sun in the east, the natural illumination of our world has been the most predictable aspect of the human experience and thus the most apt to be taken for granted. However, in some regards, human evolution can be measured against how we have used technology to transform light from the most passive element of our environment into something active, reactive and even interactive. In the process, light has become more than illumination. In this chapter, we will look at some important milestones marking light's path from human evolution to sustainable revolution, the role Philips Lighting has come to play in this transformation, challenges facing the modern lighting industry as well as the world we live in, and what the future promises. In a sense, a new revolution is at hand.

EVOLUTION: THE EARLY IMPACT OF LIGHTING

It would be hard to overestimate the fundamental importance of light as it shapes virtually everything we sense and experience. Beyond the light we need to see, the energy we use on a daily basis began with sunlight, either past or present. Fossil fuels, for instance, originated from sunlight of the distant past, just as we increasingly utilise fuels such as solar and wind power made possible by the sunlight of today.

From an anthropological standpoint, we know that human evolution was impacted when man captured and began to employ fire as a tool. Fire gave him a way to cook food, migrate to colder climes and enjoy protection from predators. However, it must not be overlooked that fire gave man the ability to use the night-time hours in ways not possible before, and these were hours free of hunting and gathering chores. This almost certainly helped form the social bonds that led to the growth of villages and later cities and urban areas.

HISTORIC MILESTONES

If we look at the history of lighting, antiquity offers important events that moved the technology of firelight forward, such as the oil lamp circa 4500 BC, or the candle 1,500 years later. However, if we look to the science of light itself, it was in 1015 that Arab scholar Ibn al-Haytham (often referred to as the father of modern optics) published his *Book of Optics*, a seven volume treatise that described light as a phenomenon of wavelengths being received by the human eye and processed by the brain. So important were al-Haytham's contributions to the study of light, vision and optics, this forms the basis for the year 2015 being selected by UNESCO as the International Year of Light, marking a millennium of advancement in the wake of al-Haytham's work.

Many others have built on these early discoveries. In 1815, French engineer and physicist Augustin-Jean Fresnel extended the wave theory of light to a larger class of optics, including his invention of a glass lens to replace the use of mirrors in lighthouses. His *Fresnel equations*, which predict the behaviour of light when moving between media of different refractive properties, continue to find numerous applications today. In 1865, with the publication of *A Dynamical Theory of the Electromagnetic Field*, Scottish scientist James Clerk Maxwell demonstrated the wave properties of electrical and magnetic fields, and proposed that electricity, magnetism and light are different manifestations of the same wave-based phenomenon. As the wave light theory became generally accepted, Albert Einstein took a slightly different approach, describing in 1905 how light can also behave as particles, or photons. Through the centuries, many of our brightest scientific minds have worked to better understand the properties of light, and these discoveries have fuelled the many technological advances to follow.

THE FIRST REVOLUTION: ELECTRIC LIGHT
IN THE HOME AND WORKPLACE

What can be considered the first revolution in lighting was the successful introduction, a little more than a century ago, of so-called 'artificial' or electric light into the home and workspace. This leap forward was made possible by the convergence of two distinct technologies: widespread electrification and the invention of a long-lasting carbon filament lamp. Building upon previous technologies such as the vacuum tube lamp and arc lamp, American inventor Thomas Edison and British physicist and chemist John Wilson Swan worked concurrently towards an incandescent bulb that utilised a carbonised paper filament burning within a glass housing, or bulb. Early carbon filament bulbs had a lifespan of 40 hours, while further refinements in filament composition increased their lifespan dramatically. Meanwhile, Edison's Pearl Street Station – the first central power plant in the United States – became operational on 4 September 1882, serving an initial load of 400 lamps and 82 customers in a one-quarter square mile area. Only two years later, the station served more than 500 customers and powered more than 10,000 lamps. The world's demand for more light and the power needed to generate it has grown exponentially in the years since.

Inspired by the fast-growing electricity industry and the promising results of his son Gerard's experiments to make reliable carbon filaments, in 1891 Frederik Philips financed the purchase of a modest factory in Eindhoven, the Netherlands. His plan was straightforward: to bring cost-effective, reliable electric incandescent light bulbs to everyone who needed them (Figure 13.1).

Philips Gloeilampen Fabrieken (Philips Incandescent Lamp Factories) began by making carbon filament lamps and quickly became one of the largest producers in Europe. From the outset, Philips was an export-oriented company. Large orders were won in Russia, including one from the Tsar to light up the Winter Palace. With developments in new lighting technology fuelling a steady programme of expansion, Philips established a research laboratory in 1914 to study physical and chemical phenomena and stimulate product innovation.

Indeed, innovation progressed beyond the limitations and energy requirements of the carbon filament. In the quest for a more efficient technology, several families of gas discharge bulb were created including the fluorescent lamp, in which an

FIGURE 13.1   The Philips Lighting workforce and factories in the early 19th century.

(a)   (b)   (c)

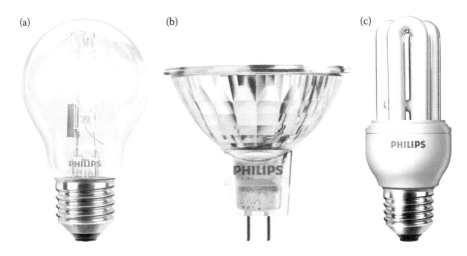

FIGURE 13.2 Conventional lighting products: (a) Incandescent lamp, (b) Halogen and (c) Compact fluorescent.

electrical current is passed through mercury vapour, exciting the molecules which emit ultraviolet light. This light is then absorbed by a phosphor coating inside the lamp which causes it to glow, or fluoresce. Fluorescent lighting came to be widely used in office and industry lighting applications because of its energy efficiency in relation to light output as compared to incandescent lamps. Compact fluorescent lamps (CFL) were later developed to replace incandescent lamps in the home. Additionally, high intensity discharge (HID) lamps, utilising an electrical arc between tungsten electrodes mounted in a gas medium, built upon the earlier technology of the simple arc lamp to deliver more visible light per unit of electricity than either incandescent or fluorescent lamps before them. This family of lamp types is used primarily for outdoor applications such as the lighting of streets, stadiums and retail locations (Figure 13.2).

Undoubtedly, the most promising technology to transform electric lighting in recent decades has been the development of the LED, or light emitting diode, a format in which voltage is applied to a two-lead semiconductor triggering the release of energy – in this case illumination – in the form of photons. The lifespan and electrical efficiency of the LED lamp is several times greater than any of the technologies that have preceded it, offering tremendous benefits in terms of energy savings. In fact, the promise and potential of global adoption of LED lighting comes at a crucial point in human history.

CHALLENGES AT HAND

When Thomas Edison threw the switch at Pearl Street Station in 1882, he directed about 600 kilowatts of electricity. Today, 7 billion citizens demand more than 20 trillion kilowatts. By 2050, with a population approaching 9.5 billion, demand could double. By that point, it is possible we could face a gap between energy supply and demand which some call the 'zone of uncertainty'. Beyond growth in population and energy demand, a variety of factors call for disruption in the lighting industry. One is increased urbanisation: today, 54% of the world's population, 3.5 billion people, live in cities. By 2050, it is projected that over two-thirds, or close to an additional 3 billion, will be urban residents. Simultaneously, an additional 3 billion will become part of the middle class, increasing their energy use accordingly. By 2030, these global trends will already lead to an estimated 35% increase in the number of light points worldwide (Figure 13.3).

Underlying these already formidable challenges of growth, the realities of climate change force an urgent collective focus on reducing energy consumption and carbon emissions. Realising the opportunity for the savings afforded by LED, on 7 December 2006, Philips took an unprecedented action in the lighting industry by calling for a global phase-out of incandescent light bulbs, the very basis and origin of the company. Governments around the world followed suit. In 2009, the European Union initiated a phase-out of incandescent bulbs which was completed by 2012. The EU later adopted the phase-out of halogen bulbs by 2016.

As Philips joined world leaders in Paris for COP21 in late 2015, the company stepped up to its role in reducing greenhouse gas emissions by committing to reduce its carbon footprint to zero by 2020. The announcement came as Eric Rondolat, Chief Executive Officer, Philips Lighting, addressed the COP21 Energy Day Summit urging leaders to set more aggressive targets to prevent climate change. 'As it stands', Rondolat said, 'we have reached the climate change checkout and all the contributions from around the world have proved insufficient to prevent a potentially catastrophic rise in global temperatures. The world must set more ambitious goals to improve energy efficiency. Faster adoption of LED lighting, and a drive to renovate

FIGURE 13.3 A surging global population and urbanisation is making improved energy efficiency a critical need.

existing city infrastructure and greater use of solar-powered LED lighting would have a huge impact'.

SECOND REVOLUTION: THE PROMISE OF LED

The promise of LED is profound, in part due to the alignment of two distinct technologies. Just as the incandescent lamp was integrated into the growing footprint of electrification more than a century ago, today the integration of connected LED technology into the Internet of Things means lighting need no longer be thought of as a matter of output but of outcomes. In the years ahead, continued advances in connected, user-centric lighting will transform the way we live, work, travel, relax, light our cities, grow our crops, heal the sick and much more.

As a value proposition, even with the formidable growth in light points by 2030, LED offers huge savings in terms of both cost of operations and toll on the environment. In 2006 – the last year the lighting market was changing at an evolutionary pace – lighting accounted for 19% of global electricity consumption. A universal switch to LED lighting would reduce this proportion to around 8%, curbing global carbon dioxide emissions by about 1,400 megatons by 2030 (Figure 13.4).

A switch to LED lighting is a necessary first step in the sequence of energy efficient innovation. LED consumes at least 40% less energy than conventional lighting. The potential savings from connected lighting – where intelligent LEDs embedded with sensors are connected wirelessly and managed remotely via the internet – can reach 80%.

Beyond energy efficiency, cost savings and reduced carbon emissions, LED offers tremendous benefits to our quality of life. LED can now be used in all applications, connected to lighting management systems and adjusted to produce new lighting experiences. In response to the global demand for more and better light, Philips has transformed its historic model as a producer of single point lighting products and become the world leader in connected lighting and systems. The following examples highlight the advantages of positioning connected LED technology within the Internet of Things while giving an exciting glimpse of things to come.

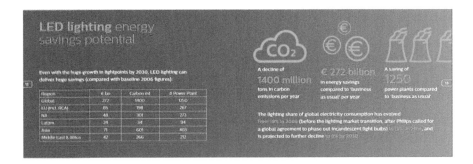

FIGURE 13.4   The savings potential from LED lighting includes benefits beyond efficiency.

CONNECTED CITIES

The City of Los Angeles's public lighting system includes almost a quarter of a million streetlights – more than any other city in the United States. Given that streets comprise fully 15% of the city's total area, it is no surprise that Los Angeles is known the world over as a driving city. However, Mayor Eric Garcetti is determined to make Los Angeles a walking city as well. In 2015, he launched a Great Streets Initiative to revitalise neighbourhoods by making the streets more pedestrian-friendly, and new street lighting technology that can ensure better and more reliable lighting operations was an important part of the plan. To support the Mayor's initiative, city officials went in search of a system that would increase street light uptimes, shorten repair cycles, and improve system monitoring and maintenance, all while minimising initial and ongoing costs (Figure 13.5).

Los Angeles's 215,000 street lights include more than 400 different styles distributed across 7,500 miles of roadway. Maintenance has traditionally depended on crews who scout the streets at night to identify outages as well as calls from citizens – as many as 40,000 per year. Connected LED street lighting is allowing officials to remotely monitor and receive automatic notification of outages and other events, while controlling each light individually or in custom groups. Monitoring and adjusting any luminaire is just a few clicks away. Custom lighting and dimming schedules can be set so that, for example, there is more light in the centre on Friday evenings and less light in the business district on weekends. With total visibility over every asset in the illumination infrastructure, officials can plan, analyse and manage the city's lighting easily and efficiently. Most importantly, straightforward remote management makes

FIGURE 13.5 The city of Los Angeles is realising many benefits from connected LED street lighting.

city lighting fully flexible, so citizens always have the illumination they need, where and when it is needed.

In 2015, Los Angeles reported an energy saving of 63% and a cost reduction of almost $9 million. The city offers a compelling model for the potential of connected street lights to deliver better, more energy-efficient light. Of approximately 300 million street lights worldwide, only about 10% are LEDs, and just 2% are connected.

CONNECTED OFFICES

The office space of the twenty-first century is rapidly evolving, and lighting plays a vital role in this transformation. A vivid example can be found in the Edge, an innovative, 40,000 square metre, multi-tenant office building in the Zuidas business district in Amsterdam. The Edge was designed to create an intuitive, comfortable and productive environment for employees that could serve as an inspiration for sustainable building designs around the world. Philips Lighting worked closely with OVG Real Estate, the building's designer, and Deloitte, its primary tenant, to deliver a connected lighting system that enhances the flexibility of the open-plan office, where workers have no fixed desk, but rather utilise a variety of shared spaces from sitting desks, standing desks, meeting rooms or private enclosures. It is what the Dutch call *het nieuwe werken:* the new way of working (Figure 13.6).

The backbone of the system is a framework of nearly 6,500 connected LED luminaires distributed throughout the building's 15 storeys. Integrated sensors in 3,000 of these luminaires work with lighting management software to form a 'digital ceiling' that captures, stores, shares and distributes information throughout the illuminated space. Through 750 Power over Ethernet switches connecting light fixtures to the

FIGURE 13.6 Connected lighting help make Amsterdam 'The Edge' the world's greenest building.

building's IT network, Ethernet cables send both power and data to the luminaires, eliminating the need for separate power cables.

Employees use the system to create a more personal working space. For example, the connected LED luminaires use visible light communications to send a code that is received by the employee's smartphone camera, registering his or her location. By using an app, the employee may then control the lighting above a specific desk, adjust the lighting and temperature in meeting rooms, and even customise the lighting for a particular activity.

Beyond increasing comfort and productivity, the system also provides building managers with real time data on operations and activities. Edge managers use the software to visualise and analyse these data, track energy consumption and streamline maintenance operations. This provides for maximum efficiency as well as a reduction of the building's CO_2 footprint. The expected savings for the Edge are €100,000 in energy costs and €1.5 million in space utilisation costs per year.

CONNECTED HOMES

Lighting is one of the main interests in the Internet of Things for the home. Philips Hue was the world's first connected lighting system for the home when it launched in 2012, and has since become the system of choice for consumers.

Connected LED technology allows light to be an interactive part of the home environment. With more than 16 million colours to enhance any atmosphere, light recipes can be designed to help users feel more energised in the morning or more relaxed after a busy day (Figure 13.7).

The Philips Hue ecosystem includes bulbs, luminaires, strips and switches which are easily controlled via apps using smartphones, tablets, wearables or even voice activation. When at home, the app can be used to control the lights in a particular

FIGURE 13.7   Connected LED technology allows for interactive home lighting.

room or the entire house, and when away from home, lights can be controlled via the Hue portal for peace of mind. By being interoperable with apps, products and platforms from other brands and developers, exciting new lighting experiences are limited only by the user's imagination.

INDOOR POSITIONING

While most smartphone users are familiar with at least one of the many global positioning apps on the market, they may not be aware of the benefits of indoor positioning systems, which can pinpoint a user's location in a store rather than on the globe. Once located, a variety of helpful information can be accessed to customise user activities. This technology is already showing great promise in the retail shopping experience, as evidenced by French shopping giant Carrefour.

Carrefour spans the globe, operating 10,800 stores in 33 countries. Always concerned with improving customer satisfaction, Carrefour conducted a research study with shoppers to learn more about their in-store experiences. The study showed that customers wanted to be able to find products more easily.

Embracing the use of indoor positioning technology, Carrefour installed a Philips connected lighting system with LED-based indoor positioning during the renovation in 2015 of its hypermarket in the Euralille mall in Lille – one of the top ten shopping malls in Europe. The installation's 800 LED fixtures use visible light communications to send a unique code that is invisible to the human eye, but readily detected by a smartphone camera. The smartphone reads this code and pinpoints exactly where the visitor is standing in the store, to a distance of under half a metre. After downloading an app and selecting promotions from a catalogue, a shopper becomes oriented with a map and may start navigating to the list of selected items. As one moves around the store, the app can also show promotions on nearby items, so a deal is never missed.

Owing to its energy efficiency, the system was a natural fit for Carrefour. The company has announced its intention to reduce overall energy consumption by 30% and CO_2 emissions by 40% by 2020. Connected LED lighting will help achieve this goal, reducing energy consumption by 50%.

ELIMINATING LIGHT POVERTY

As governments across the world look for ways to curb carbon emissions, improving energy efficiency is an urgent priority. In developing regions, solar lighting has given us huge opportunities to 'leapfrog' outdated technology.

New projects in India vividly demonstrate this potential. The Indian states of Uttar Pradesh and Manipur, for example, have begun installation of more than 76,000 solar LED street lamps in rural communities. Solar LED is less expensive, more environmentally friendly and eliminates the noxious fumes from alternatives such as kerosene (Figure 13.8).

Today, almost 300 million Indians depend on wood fires or kerosene for their lighting. Worldwide, about 1.1 billion people are trapped in light poverty, denied access to reliable electricity. Clean solar light is a simple, fast solution to this injustice. Dramatic improvements can be achieved simply by making better use of

FIGURE 13.8   Solar LED is one way of helping to tackle 'light poverty'.

technology that is already available. The new installations in Uttar Pradesh will light more than 800 villages and small towns, following a tender process focused on new and renewable energy sources.

Philips is working with Uttar Pradesh New and Renewable Energy Development Agency (UPNEDA) and Manipur Renewable Energy Development Agency (MANIREDA) to supply LED luminaires powered by chargers that deliver energy efficiency of more than 96%. The housings are designed for harsh conditions, notably extreme temperatures and exposure to water and dust.

Off-grid solar lighting is an important tool in driving down carbon emissions and accelerating global development. We do not have to wait for answers or new inventions. The technology we need is already transforming the lives of off-grid and urban communities in India.

SOMETHING MORE THAN ILLUMINATION

From the capturing of fire as a tool, through technical advances that are transforming electric light into something more than illumination, lighting has undergone a dramatic evolution and seen equally dramatic periods of revolution. In fact, revolution is again in the air. In the years to come, we will see the complete disappearance of the nineteenth century incandescent light bulb – the end of the first mass electrical appliance – to be replaced by twenty-first century connected LED lighting systems and technology. Through ceaseless innovation and a commitment to the opportunities these breakthroughs afford, a new revolution is at hand: an era of more and better light – light that provides for a more sustainable world.

14 Mediaeval Arab Achievements in Optics

Sameen Ahmed Khan

CONTENTS

Introduction 193
History of Ancient Optics 193
Mediaeval Arab Achievements in Optics 194
Synchrotron Light and International Science Collaborations 198
Concluding Remarks 198
Acknowledgements 199
References 199

INTRODUCTION

When we browse through any physics or optics textbook, we invariably come across 'Snell's law' of refraction of light named after the Dutch scientist, Willebrord Snellius (1580–1626). The law was stated by Snell in a manuscript dated 1621. The same law is also attributed to Thomas Herriot (1560–1621), René Descartes (1596–1650) and Fermat (1601–1665) [1]. In 1990, mathematician and science historian Roshdi Rashed brought to light the discovery that the refraction of light was known to and written by the Mediaeval Arab scholar Ibn Sahl (940–1000) [2,3]. In this article, we shall describe this historical discovery based on Rashed's examination of Ibn Sahl's treatise dated 984, which is more than six centuries earlier than Snell. The rest of the article is organised as follows. In section 'History of Ancient Optics', we shall briefly recollect the outline from the history of ancient optics, in particular Greek optics. In section 'Mediaeval Arab Achievements in Optics', we shall look at how Greek optics reached the Arab lands. Then we shall review the Mediaeval Arab achievements in optics and its influence on the European sciences. In section 'Synchrotron Light and International Science Collaborations', we look ahead and examine the ways in which the light sciences can provide a platform for international science collaborations in the Middle East and elsewhere. Section 'Concluding Remarks' has our concluding remarks.

HISTORY OF ANCIENT OPTICS

Humans were fascinated by light since the very beginning of civilisation. This can be ascertained from the archaeological findings along with ancient drawings and writings across the civilisations. On a technical front, lenses made from quartz can be dated to be about 3000 years old. These are the Nimrud/Layard lenses discovered by Austen Henry Layard in 1850, during the excavations of the Assyrian Palace of Nimrud in Iraq. Similar lenses were also unearthed at other archaeological sites, including Babylon, Egypt and Greece. They might have been used as magnifying glasses and possibly for lighting fires. Other archaeological studies indicate that Greeks and Romans used lenses made by filling glass spheres and possibly other geometries.

Apart from archaeological artefacts, there are also works supporting the fact that the earlier civilisations had made attempts to understand optical phenomena as a discipline in its own right. For instance, the Greeks had three distinct theories of vision. The *extramission* theories resorted to illuminating particles released from the eyes. The *intromission* theories assumed that the object released the particles, which reached the observer's eyes. The third theory used a combination of both. The supporters of the intromission theory include Aristotle. Euclid used the extramission theory to develop a geometrical description. The geometric description was detailed and useful but was obviously based on a wrong assumption. The optical phenomenon could not be fully explained using the intromission theories. Euclid's work is important as it developed a geometric theory of optics. Euclid formulated the laws of *reflection* of light, which could successfully describe the images in the case of reflection. The geometric theories of Euclid influenced the studies of *refraction* carried out by the astronomer Claudius Ptolemy.

The intromission and extramission theories addressed the nature of light and the mathematical aspects, respectively. The other important approach to vision and the eye was from a medical perspective. Galen of Pergamon was a distinguished physician and surgeon in the Roman Empire during the second century. Galen is famous for his description of the eye. He could identify the *crystalline humour* as the lens situated at the centre of the eye. He further stated that the lens was the primary part of the visual system as it was sensitive to light. Galen believed in the intromission theories.

MEDIAEVAL ARAB ACHIEVEMENTS IN OPTICS

The Mediaeval Arabs were great patrons of science. They took a keen interest in collecting the books from various civilisations. Since, the books were in different languages, they started the process of translating these collected books into Arabic. The mammoth task of translations lasted for over two centuries (eighth and ninth centuries CE) under royal patronage and enabled the preservation of scientific texts from diverse sources. Many of the original texts in Greek, Pahlavi, Sanskrit, Syriac and other languages have been lost and we now only have their Arabic translations. This remarkable enterprise of preservation of the ancient bodies of knowledge has earned the Mediaeval Arabs an elite status [4]. The translation activities, along with original contributions, created a unique scientific environment in the Arab lands during the eighth to thirteenth centuries. Historians call it the *Islamic Golden Age of Science*. The translations included the Greek classics such as Ptolemy's *Optics* and Apollonius' *Conic Sections*. Anthemius of Tralles (474–558 CE) had made detailed descriptions of the parabolic mirrors and their use in burning instruments. Arabs were aware of Tralles's works. Thus, the Arabs had inherited and importantly preserved the bodies of knowledge from other civilisations.

It is to be noted that the task of translations was *not* done by linguistics but by the subject experts [4]. The task of translating Ptolemy's *Optics* was done by Abu Said al-Ala Ibn Sahl (940–1000). Ibn Sahl translated other Greek works as well. In the previous section, we noted that Euclid had developed the theory of reflection. Ibn Sahl made several contributions to optics and in 984 wrote the book *Kitāb al-Parraqāt (On the Burning Instruments)*. The law of *refraction* of light is found in this book. This book of Ibn Sahl had been cited extensively by others and, of course, by his illustrious student Ibn al-Haytham [5–9]. Manuscripts bearing the title *Kitāb al-Parraqāt* were available in the libraries in Damascus, Syria and Tehran, Iran. Till 1990, it was believed that the manuscripts were copies of the same book. In this year, Roshdi Rashed started examining these two manuscripts. On the basis of the contents and the general structure, Roshdi Rashed could conclude that both the manuscripts were actually parts of *Kitāb al-Parraqāt*. Roshdi Rashed went a step further and translated the reassembled book [2,3,10,11]. This piece of work by Roshdi Rashed clearly indicated that Ibn Sahl had penned down the *law of refraction of light* over half a millennium before his European counterparts. For this major historical discovery, Roshdi Rashed earned several major awards including: the TWAS Prize in 1990 (Third World Academy of Sciences, now renamed as The World Academy of Sciences, in Trieste, Italy); UNESCO's Avicenna Gold Medal in 1999; the CNRS Medal in 2001 (French National Centre for Scientific Research); and the KFIP in 2007 (King Faisal International Prize). The historical discoveries of Roshdi Rashed have enabled us to have a much better picture of the Mediaeval Arab contributions to optics as well as other sciences [12].

Now we shall have a closer look at Ibn Sahl's contributions. Ibn Sahl's book is comprehensive as it covers the theoretical and experimental aspects in detail. The book covers the ways of drawing conic sections with the aid of mechanical devices. For instance, the ellipse can be drawn with a string and two holders for the two foci. Other conic sections require more elaborate means. As stated earlier, the book contains the law of refraction of light. Figure 14.1 is reproduced from Ibn Sahl's treatise. We shall examine it in detail.

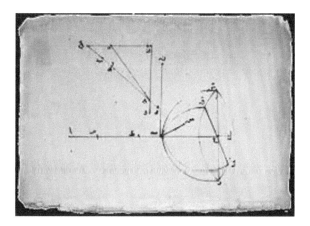

FIGURE 14.1   Ibn Sahl's diagrams for refraction of light and the plano-convex lens. (Milli MS 867, folio 7r; Milli Library Tehran, Iran.)

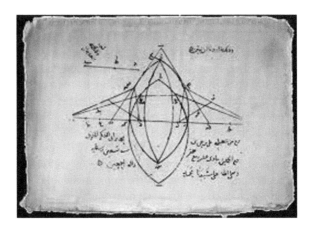

FIGURE 14.2 Ibn Sahl's diagram for a biconvex hyperbolic lens. (Milli MS 867, folio 26r; Milli Library Tehran, Iran.)

In his book, Ibn Sahl also covers a variety of geometries for lenses (hyperbolic biconvex and hyperbolic plano-convex) and mirrors (ellipsoidal and parabolic), respectively. Figure 14.2 has the biconvex hyperbolic lens (p. 467 in Reference 2). Ibn Sahl's book points to the fact that he and others were actively doing research on the refraction of light, towards the end of the tenth century [2,3,5]. It also points to the fact that the works reported in Ibn Sahl's book were firmly based on experiments and geometric techniques [1–3,5].

Contemporary scholars have made a detailed examination of Ibn Sahl's diagrams [2,10,11]. Figure 14.3 has the reconstructed schematic figure of Ibn Sahl's diagram. In Figure 14.3, the vertical line OB separates two media with refractive indices n_1 and n_2, respectively. AB is the incident ray of light, while BC is the refracted ray, when it enters the second medium. The dashed line BD represents

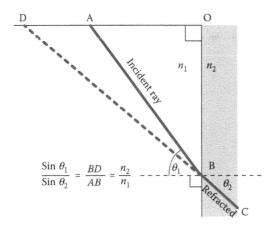

FIGURE 14.3 Ibn Sahl's diagram for refraction reconstructed.

the extension of the refracted ray path BC. The incidence angle θ_1 is the angle between the incident ray AB and the normal to the boundary surface OB. The angle of refraction θ_2 is the angle between the refracted ray BC and the normal to OB. Ibn Sahl found that the angle OAB is equal to θ_1 and the angle ODB is equal to θ_2. Therefore, we have the law of refraction, $n_1 \sin \theta_1 = n_2 \sin \theta_2$, which means that the ratio of the sines of angle of incidence and angle of refraction is a constant for a given pair of media. Additional technical details about the diagrams are available in References 2, 10 and 11.

Ibn Sahl's illustrious student Abu Ali al-Hasan Ibn al-Haytham is perhaps more widely known than his master! Ibn al-Haytham was born in Bassorah (Basra), Iraq in 965 and died in Cairo, Egypt in 1040. The works of Ibn al-Haytham were frequently cited by his contemporaries and the later generation of mediaeval scholars. The Latin and other language translations of his works were also used and cited widely. The bibliographies generate a list of 96 manuscripts authored by Ibn al-Haytham. Of these, fourteen are on optics. The most famous is the seven volume *Kitāb al-Manazir* (*Book of Optics*). Numerous optics-related anniversaries were marked during the IYL2015. Ibn al-Haytham's *Kitāb al-Manazir* provided the millennium anniversary! Ibn al-Haytham remained the key figure in the IYL2015 documents and the year-long celebrations of light. The detailed coverage of the voluminous contributions of Ibn al-Haytham is beyond the scope of this article. Ibn al-Haytham's work included curved mirrors; spherical aberrations; and double refraction in a sphere. More importantly, he made some general statements such as: *light travels in straight lines*; *the speed of light differs from material to material*; and *light has a finite speed*. He further stated that *light takes a path which is easier and quicker*, which we now know as Fermat's principle of least time [13,14]. The opto-mechanical analogy had a simple beginning in the works of Ibn al-Haytham much before that of Descartes in the seventeenth century [15]. Ibn al-Haytham was well equipped with mathematics and the aforementioned statements or principles. These enabled him to handle several challenging problems of his time including: formations of rainbows, dispersion of light into constituent colours and optical illusions. In passing, we note that it was only in 1833 that Hamilton laid a rigorous mathematical foundation to the opto-mechanical analogy [16]. In the 1920s, the opto-mechanical analogy had a decisive influence on the development of quantum mechanics. In recent decades, Hamilton's opto-mechanical analogy was extended into the wavelength-dependent regime [17–27].

The Latin translations of *Kitāb al-Manazir* can be traced to the 1230s. There were several translations in Latin, Hebrew and several European languages. The European writings on optics have ample citation to the original Arabic manuscript and the translated versions. From these citations, it is evident that all the scientists of the European Renaissance had access to the reservoir of knowledge preserved and created by the Mediaeval scholars in the Arab lands [28–34].

Our narration was essentially confined to Ibn Sahl and Ibn al-Haytham. Other Medieval scholars in the field of optics include: Yaqub Ibn Ishaq Ibn Sabah al-Kindi (801–873); Hunayn ibn Ishaq (809–873); Abu Sahl Waijan Ibn Rustam al-Quhi (940–1000); Qutb al-Din al-Shirazi (1236–1311); and Kamal al-Din al-Farisi (1267–1319). The detailed and comprehensive accounts are to be found in References 31 and 32.

SYNCHROTRON LIGHT AND INTERNATIONAL SCIENCE COLLABORATIONS

In the preceding sections, we have noted Mediaeval Arab contributions to optics. However, today, the Arab lands are far behind in their scientific output [35]. It is interesting to note that light sciences can be a possible source of scientific revival in the Middle East. The light sciences and associated technologies are impacting every facet of human existence. In Europe, after the devastation of two world wars, the warring nations had come together through the platform of scientific collaborations. The optics-related large-scale collaboration is the European Synchrotron Radiation Facility (ESRF). Let us recall that synchrotron light is a special form of radiation much more powerful than the conventional x-rays and lasers. Synchrotron light enables diverse applications in basic and applied sciences, which are not possible with traditional x-rays and lasers. However, the synchrotron light sources require billions of dollars to build and an optimum technical expertise. These aspects have led to international collaborations in building and running synchrotron light facilities. ESRF is jointly supported by 21 countries [36]. The Middle East's first synchrotron light source is located in Jordan. It is known by the magic acronym SESAME: Synchrotron-light for Experimental Science and Applications in the Middle East. Presently, SESAME has nine member countries and the list is expected to grow [37–40]. The SESAME facility was created by relocating a German synchrotron. This happened as Germany decided to have a larger facility. It is to be noted that SESAME is *not* the first relocated synchrotron. In 1996, Japan donated a synchrotron to Thailand, heralding the era of relocated synchrotrons [41,42]. The Siam Photon Source is Thailand's first synchrotron light facility and is serving scientists throughout Southeast Asia. The other upcoming synchrotron in the Middle Eastern region is the Iranian Light Source Facility [43–45].

Barring the inhabitable Antarctica, the continent of Africa is the only region without a synchrotron. The idea of an African synchrotron has been under discussion for over a decade [46–50]. Siam and SESAME are unique facilities as they are being built by relocating the very generously donated synchrotrons. The *World Synchrotron Map* has wide gaps [51]. These can be filled by the upcoming facilities and by planning synchrotrons in the under-represented regions. These, when built, will immensely benefit the scientific community in the concerned regions by enhancing international cooperation and providing them the latest technological expertise, thus providing an excellent opportunity to scientists to be catapulted to the forefront of the latest technological expertise.

CONCLUDING REMARKS

We briefly noted the interest of optics in the ancient civilisations, particularly in the Greek tradition. The Greek works reached the Arab lands through acquisitions. We also noted how the Arab scholars preserved the ancient contributions to sciences. These works, preserved in Arabic, were later used by the Europeans, who also followed the path of translations accompanied by original contributions. On the basis of the recent analysis of Ibn Sahl's work by the mathematician and science historian Roshdi Rashed, there is *no* doubt that Ibn Sahl correctly understood the law of refraction and also used it to explain lenses of complex geometries ([2]; the note by Bettini in [9] and [10,11]). Kurt Bernardo Wolf states

> We feel obliged to put a new name to the sine law of refraction that was written some 650 years before Willebrord Snell found it experimentally and René Descartes by contrived (and incorrect) theoretical reasoning [10].

In the just concluded United Nations designated *International Year of Light and Light-based Technologies*, Ibn al-Haytham received good recognition [9,52–56]. This would have been an ideal occasion to give due recognition to Ibn Sahl as the originator of the law of refraction of light.

We briefly noted the important place enjoyed by light sciences in the arena of international collaborations. We cited the examples of the ESRF, SIAM and SESAME synchrotrons. Such collaborations can be replicated in other regions, particularly in Africa. One of the long-term legacies of the International Year of Light is the newly created *Ibn al Haytham LHiSA International Society*, where LHiSA stands for Light: History, Science and Applications [54,57]. LHiSA is the brainchild of the Optical scientist Prof. Azzedine Boudrioua from the University of Paris. LHiSA is an excellent platform to bring about international collaboration in light sciences and other disciplines [58,59]. Such collaboration would be the turning of a new leaf in the Middle East, which was a fertile land of science a few centuries back [35]. Undoubtedly, we need such collaborations in Africa, South America and the 'developing' regions of the world [60].

ACKNOWLEDGEMENTS

I am very thankful to Prof. Azzedine Boudrioua for extending an invitation to me for the Conference. I profusely thank the UNESCO for providing me complete financial support towards air travel and local hospitality. In the absence of such support, my participation would have been difficult. I greatly appreciate the warm hospitality, which I enjoyed throughout my stay for the *Landmark September Event*.

REFERENCES

1. A Kwan, J Dudley and E Lantz, Who really discovered Snell's Law? *Physics World*, 15 (4), 64, April 2002. http://dx.doi.org/10.1088/2058-7058/15/4/44
2. R Rashed, A pioneer in anaclastics: Ibn Sahl on burning mirrors and lenses, *ISIS*, 81, 464–491, 1990 (The University of Chicago Press). http://www.jstor.org/stable/233423 (accessed date: 17/05/2017).
3. R Rashed, Géométrie et Dioptrique au Xe siècle: Ibn Sahl, al-Quhî et Ibn al-Haytham, *Collection Sciences et Philosophie Arabes, Textes et Études, Les Belles Lettres*, Paris, France, 1993. http://www.lesbelleslettres.com/ (accessed date: 17/05/2017).
4. Z Virk, Europe's debt to the Islamic world, *Review of Religions*, 98 (7), 36–52, 2003.
5. R Rashed, A polymath in the 10th century, *Science*, 297, 773, 2002. http://dx.doi.org/10.1126/science.1074591
6. D C Lindberg, Medieval Islamic achievement in optics, *Optics & Photonics News*, 14 (7), 30–35, 2003; Akhlesh Lakhtakia, Optics in medieval Islam, *Optics & Photonics News*, 14 (9), 6, 2003.

7. C M Falco, Ibn al-Haytham and the Origins of the Modern Image Analysis, Proceedings of the International Conference in Information Sciences, Signal Processing and its Applications (12–15 February 2007, Sharjah, United Arab Emirates).

8. D Hockney, J T Spike, S Grundy, R Lapucci, F Camerota, C M Falco and C Pernechele, Ibn al-Haytham's Contributions to Optics, Art, and Visual Literacy, in Painted Optics Symposium, (7–9 September 2008, Florence, Italy).

9. S A Khan, Medieval Islamic Achievements in Optics, *Il Nuovo Saggiatore*, 31 (1–2), 36–45 (January–February 2015) (Società Italiana di Fisica, the Italian Physical Society). http://prometeo.sif.it/papers/online/sag/031/01-02/pdf/06-percorsi.pdf (accessed date: 17/05/2017).

10. K B Wolf and G Krötzch, Geometry and dynamics in refracting systems, *European Journal of Physics*, 16 (1), 14–20, 1995.

11. K B Wolf, *Geometric Optics on Phase Space*, pp. 8–9 and pp. 57–58, (Springer, Berlin, Germany), 2004.

12. Sameen Ahmed Khan, Arab origins of the discovery of the refraction of light; Roshdi Hifni Rashed Awarded the 2007 King Faisal International Prize, *Optics & Photonics News*, 18 (10), 22–23, 2007. http://www.osa-opn.org/Content/ViewFile.aspx?id=10890 (accessed date: 17/05/2017).

13. Abdus Salam, Scientific thinking: Between secularisation and the transcendent, in *Ideals and Realities*, third edition, pp. 283, Editors: C H Lai and A Kidwai (World Scientific, Singapore), 1989.

14. Abdus Salam, The Gulf university and the science in the Arab-Islamic commonwealth, in *Renaissance of Sciences in Islamic Countries*, pp. 31, Editors: H R Dalafi and M H, A Hassan (World Scientific, Singapore), 1994.

15. D Ambrosini, A Ponticiello, G Schirripa Spagnolo, R Borghi and F Gori, Bouncing light beams and the Hamiltonian analogy, *European Journal of Physics*, 18, 284–289, 1997.

16. M Born and E Wolf, *Principles of Optics* (Cambridge University, London, UK), 1999.

17. R Jagannathan, R Simon, E C G Sudarshan and N Mukunda, Quantum theory of magnetic electron lenses based on the Dirac equation, *Physics Letters A* 134, 457–464, 1989. http://dx.doi.org/10.1016/0375-9601(89)90685-3

18. R Jagannathan, Quantum theory of electron lenses based on the Dirac equation, *Physical Review*, A42, 6674–6689, 1990. http://dx.doi.org/10.1103/PhysRevA.42.6674

19. M Conte, R Jagannathan, S A Khan and M Pusterla, Beam optics of the Dirac particle with anomalous magnetic moment, *Particle Accelerators*, 56, 99–126, 1996. http://cds.cern.ch/record/307931/files/p99.pdf (accessed date: 17/05/2017).

20. R Jagannathan and S A Khan, Quantum theory of the optics of charged particles, Chapter-4 in *Advances in Imaging and Electron Physics*, Editors: P W Hawkes, B Kazan and T Mulvey, 97, 257–358 (Academic Press, San Diego), 1996. http://dx.doi.org/10.1016/S1076-5670(08)70096-X

21. S A Khan, Quantum theory of charged-particle beam optics, University of Madras, Chennai, India, 1997 (PhD thesis). Complete thesis available from Dspace of IMSc Library, The Institute of Mathematical Sciences (IMSc), Chennai, India, where the doctoral research was done, http://www.imsc.res.in/xmlui/handle/123456789/75?show=full and http://www.imsc.res.in/xmlui/ (accessed date: 17/05/2017).

22. R Jagannathan and S A Khan, Quantum mechanics of accelerator optics, *ICFA Beam Dynamics Newsletter*, 13, 21–27, 1997 (International Committee for Future Accelerators). http://icfa-usa.jlab.org/archive/newsletter/icfa_bd_nl_13.pdf (accessed date: 17/05/2017).

23. Sameen Ahmed Khan, Analogies between light optics and charged-particle optics, arXiv:physics/0210028 (physics.optics); *ICFA Beam Dynamics Newsletter*, 27, 42–48, 2002 (International Committee for Future Accelerators). http://icfa-usa.jlab.org/archive/newsletter/icfa_bd_nl_27.pdf (accessed date: 17/05/2017).

24. Sameen Ahmed Khan, Wavelength-dependent modifications in Helmholtz Optics, in *International Journal of Theoretical Physics*, 44 (1), 95–125, 2005. http://dx.doi.org/10.1007/s10773-005-1488-0

25. Sameen Ahmed Khan, The Foldy-Wouthuysen Transformation Technique in Optics, Chapter-2 in *Advances in Imaging and Electron Physics*, Editor: P W Hawkes, 152, 49–78 (Elsevier, Amsterdam, Netherlands), 2008. http://dx.doi.org/10.1016/S1076-5670(08)00602-2

26. Sameen Ahmed Khan, Aberrations in maxwell optics, *Optik – International Journal for Light and Electron Optics*, 125 (3), 968–978, 2014. http://dx.doi.org/10.1016/j.ijleo.2013.07.097

27. Sameen Ahmed Khan, Passage from scalar to vector optics and the Mukunda-Simon-Sudarshan theory for paraxial systems, *Journal of Modern Optics*, 63 (17), 1652–1660, 2016. http://dx.doi.org/10.1080/09500340.2016.1164257

28. A I Sabra, ed., *The Optics of Ibn al-Haytham, Books I-II-III: On Direct Vision*. The Arabic Text, Edited with Introduction, Arabic-Latin Glossaries and Concordance Tables (National Council for Culture, Arts and Letters, Kuwait), 1983.

29. A I Sabra, trans., *The Optics of Ibn al-Haytham. Books I-II-III: On Direct Vision. English Translation and Commentary* (2 volumes set), Studies of the Warburg Institute (The Warburg Institute, University of London, London), 1989.

30. A I Sabra, ed., *The Optics of Ibn al-Haytham*. Edition of the Arabic Text of Books IV-V: On Reflection and Images Seen by Reflection (2 volumes) (The National Council for Culture, Arts and Letters, Kuwait), 2002.

31. Jim Al-Khalili, *Pathfinders: The Golden Age of Arabic Science* (Penguin Books, London, UK), 2010.

32. Mourad Zghal, Hamid-Eddine Bouali, Zohra Ben Lakhdar and Habib Hamam, The first steps for learning optics: Ibn Sahl's, al-Haytham's and Young's works on refraction as typical examples, Proceedings of The Education and Training in Optics and Photonics Conference (ETOP) 2007 (3–5 June 2007, Ottawa, Canada), Education and Training in Optics and Photonics. Optical Society of America, 2007. http://spie.org/Documents/ETOP/2007/etop07fundamentalsII.pdf (accessed date: 17/05/2017).

33. Sameen Ahmed Khan, Medieval Arab contributions to optics, *Digest of Middle East Studies (DOMES)*, 25 (1), 19–35, 2016. http://dx.doi.org/10.1111/dome.12065

34. Sameen Ahmed Khan, International year of light and history of optics, Chapter-1 in *Advances in Photonics Engineering, Nanophotonics and Biophotonics*, Editor: Tanya Scott, pp. 1–56 (Nova Science Publishers, New York), 2016. http://isbn.nu/9781634844987 and https://www.novapublishers.com/ (accessed date: 17/05/2017).

35. Nidhal Guessoum and Athar Osama, Institutions: Revive universities of the Muslim world, *Nature* 526 (7575), 634–636, 2015. http://dx.doi.org/10.1038/526634a

36. ESRF Website: http://www.esrf.fr/ (accessed date: 17/05/2017).

37. SESAME Website: http://www.sesame.org.jo/ (accessed date: 17/05/2017).

38. Sameen Ahmed Khan, The Middle East synchrotron laboratory and India, *Current Science*, 80 (2), 130–132, 2001. http://www.iisc.ernet.in/currsci/jan252001/130.pdf (accessed date: 17/05/2017).

39. Azher Majid Siddiqui and Sameen Ahmed Khan, SESAME, the first International Science Centre in the Middle East: A step towards the Renaissance of science in the Islamic countries, *Journal of Islamic Science*, 17 (1–2), 9–34, 2001 (Muslim Association for the Advancement of Science, Aligarh, India).

40. Sameen Ahmed Khan, The Middle East Synchrotron Facility can bring regional cooperation, *Digest of Middle East Studies (DOMES)*, 11 (2), 57–71, 2002. http://dx.doi.org/10.1111/j.1949-3606.2002.tb00457.x

41. T Feder, Thailand recycles Japanese synchrotron light source, *Physics Today*, 52 (8), 55, 1999; AAPPS Bulletin, 9, 44–45, 1999.

42. Siam Website: http://www.slri.or.th/ (accessed date: 17/05/2017).

43. Iranian Light Source Website: http://ilsf.ipm.ac.ir/ (accessed date: 17/05/2017).

44. Sameen Ahmed Khan, X-Rays to synchrotrons and the International Year of Light, *IRPS Bulletin*, 28 (1), 9–13, 2014 (International Radiation Physics Society). http://www.canberra.edu.au/irps and http://radiationphysics.org/ (accessed date: 17/05/2017).

45. Sameen Ahmed Khan, Particle accelerators and the International Year of Light, *ICFA Beam Dynamics Newsletter*, 63, 9–15, 2014 (International Committee for Future Accelerators). http://icfa-usa.jlab.org/archive/newsletter/icfa_bd_nl_63.pdf (accessed date: 17/05/2017).

46. K Jackson, S A Khan, A Kebde and A M Siddiqui, SESAME the Nearest Synchrotron Radiation Facility to Africa, at the website of the *African Synchrotron User Forum*. http://sirius-c.ncat.edu/asn/Africasynchrotron/index.html (20 March 2003).

47. Sameen Ahmed Khan, The Middle East synchrotron is launched; Armenian synchrotron; To launch the African Synchrotron Programme, *AAPPS Bulletin*, 13 (2), 35–36, 2003 (Association of Asia Pacific Physical Societies). http://aappsbulletin.org/ (accessed date: 17/05/2017).

48. Sameen Ahmed Khan, Ground breaking for the Middle East synchrotron; Armenian synchrotron; Time to launch the African Synchrotron Research Programme, *ICFA Beam Dynamics Newsletter*, 30, 88–89, 2003 (International Committee for Future Accelerators). http://icfa-usa.jlab.org/archive/newsletter/icfa_bd_nl_30.pdf (accessed date: 17/05/2017).

49. T Feder, Momentum grows for African light source, *Physics Today*, 69 (1), 31, 2016. http://dx.doi.org/10.1063/PT.3.3050

50. Sekazi Mtingwa, A shining light for African science, *Physics World*, 29 (1) 17, 2016. http://dx.doi.org/10.1088/2058-7058/29/1/24

51. Light Sources of the World: http://www.lightsources.org/regions

52. Sameen Ahmed Khan, 2015 declared the International Year of Light and Light-based Technologies, *Current Science*, 106 (4), 501, 2014. http://www.currentscience.ac.in/Volumes/106/04/0501.pdf (accessed date: 17/05/2017).

53. Sameen Ahmed Khan, *International Year of Light and Light-based Technologies*, 96 pages (LAP LAMBERT Academic Publishing, Germany), 2015. http://www.lap-publishing.com/ and http://isbn.nu/9783659764820/ (accessed date: 17/05/2017).

54. Sameen Ahmed Khan, The International year of light and light-based technologies, (*Report of the Conference, The Islamic Golden Age of Science for today's Knowledge-based Society: The Ibn Al-Haytham Example*, UNESCO Headquarters, Paris, France, 14–15 September 2015), *American Journal of Islamic Social Sciences*, 33 (1), 160–163, 2016.

55. J R González and J Dudley, Inspired by light: Close of the International Year of Light, *Europhysics News*, 47 (2), 6–7, 2016. http://www.europhysicsnews.org/articles/epn/pdf/2016/02/epn2016-47-2.pdf (accessed date: 17/05/2017).

56. S A Khan, Reflecting on the International year of light and light-based technologies, *Current Science,* 111 (4), 627–631, 2016. http://dx.doi.org/10.18520/cs/v111/i4/627-631

57. Ibn al-Haytham LHiSA International Society: https://www.ibnalhaytham-lhisa.com/ (accessed date: 17/05/2017).

58. Sameen Ahmed Khan, Need to create Regional synchrotron radiation facilities (RSRF), *IRPS Bulletin*, 17 (2), 7–13, 2003 (International Radiation Physics Society). http://www.canberra.edu.au/irps/bulletin/docs/Vol-28-No-1-April,-2014.pdf (accessed date: 17/05/2017).

59. Sameen Ahmed Khan, When will there be an Asian Accelerator Laboratory? *ICFA Beam Dynamics Newsletter*, 28, 49–54, 2002 (International Committee for Future Accelerators). http://icfa-usa.jlab.org/archive/newsletter/icfa_bd_nl_28.pdf (accessed date: 17/05/2017).

60. A M Siddiqui and S A Khan, Need to create International science centres in Arab countries, Chapter-15 in *Light-Based Science: Technology and Sustainable Development, Proceedings of The Islamic Golden Age of Science for today's Knowledge-based Society: The Ibn Al-Haytham Example* (14–15 September 2015, UNESCO Headquarters, Paris, France), Editors: Azzedine Boudrioua, Roshdi Rashed and Vasudevan Lakshminarayanan, pp. 207–220 (CRC Press, Taylor & Francis, UK), 2017. http://isbn.nu/9781498779388/ and https://www.crcpress.com/Light-Based-Science-Technology-and-Sustainable-Development/Lakshminarayanan-Roshdi-Azzedine/9781498779388

Section III

Optics and Photonics in the Arab and Islamic World, Education and Investment in Light Sciences and Technology

The last section of this book gives a highlight of optics and photonics in the Arab world. Education and investment in light sciences are crucial and may contribute to the scientific and technological development of the Arab world.

Azher Majid Siddiqui and Sameen Ahmed Khan address the theme of international science collaborations, in the light of examples of the institutions from Europe, which were built after the Second World War. These international institutions devoted to science played a pivotal role in the rebuilding of Europe devastated by the wars. CERN, the European Laboratory for Particle Physics in Geneva, Switzerland is a prime example of international collaboration, which now has 21 member countries. International science collaborations served as vehicles of communication and peace during the cold war. Synchrotron-light for Experimental Science and Applications in the Middle East (SESAME) in Jordan is modelled after CERN. Both CERN and SESAME are under the auspices of UNESCO. The SESAME facility is sure to advance science and promote cooperation in the region and beyond. The article discusses the need to create international science centres in the Arab countries. The proposed centres can be modelled after the European institutions keeping in mind the local perspective. Such centres, when created, shall lead to the renaissance of sciences in the region.

Gihan Kamel turns the wheel of history, seeking the initial steps towards the understanding of the optical systems based on the invaluable contribution of the Muslim and Arab Scholar, Ibn al-Haytham, and presents the enthusiastic success to develop modern eyes capable of resolving the matter down to its atoms, known as Synchrotron light sources. These are super microscopes with different experimental techniques known as beamlines; they are powerful enough to reveal the finest details about the unknowns of physics, chemistry, biology, pharmacy and biomedicine, as well as materials science, environment, art restoration, cultural heritage and some other arenas. The SESAME is one of those eyes, and is the first facility of its kind in the Middle East constructed in Jordan.

For Abdulaziz Alswailem, communication is the tool for sharing, acquiring and exploring ideas and knowledge. Different languages are used for communicating ideas and literary works including science. Translation plays a very important role across all different languages including the Arabic language. Saudi Arabia through the King Abdulaziz City for Science and Technology is an example of how to have a vision of building a knowledge-based society and economy and a globally competitive national system for science, technology and innovation. The strategic direction is for the Kingdom of Saudi Arabia to reach the ranks of developed countries in science, technology and innovation by the year 2030 through projects on scientific translation and scientific awareness outreach programmes, thus making science available for all.

Marie Abboud presents a review of some Lebanese contributions to the field of optics and photonics: from the Ksara Observatory set up in the early twentieth century to the countless inventions and patents of Hassan Kamel al-Sabbah the Oriental Edison, and up to the contemporary research performed in the fields of classical optics, modern optics, photonics, emerging fields and all related applications. She lights the Lebanese way in this research area by highlighting progress, and emphasising strengths and challenges through facts and figures.

15 Need to Create International Science Centres in Arab Countries

Azher Majid Siddiqui and Sameen Ahmed Khan

CONTENTS

Introduction 207
Reconstruction of European Science after World War II 209
SESAME Project 211
Need to Create Regional Science Centres.. 211
International Year of Light 213
Concluding Remarks... 215
Acknowledgements... 217
References... 217

INTRODUCTION

We shall concentrate on the focal theme of international science collaboration, urging the pressing need to create regional science centres in developing countries, in particular the Arab and Muslim countries.

Technology and science are interdependent as one leads to the other. Science impacts technology and technology is required for exploring nature, leading to new science. Both science and technology are crucial for the social well-being and sustainable economic development. New technologies and approaches are required to combat problems hindering social and economic development in 'developing countries'. The problems include the lack of basic necessities such as potable water, nutritious food and basic infrastructure including electricity, energy sources and healthcare. Then there is the formidable challenge to provide basic education to the budding next generation. There needs to be optimal 'capacity in science and technology' in order to have viable solutions to the aforementioned basic necessities and challenges. Science and technology lead to improvements of products and basic services such as communications and transport. In turn, better services lead to better enterprises and increased competitiveness of countries. Hence, it is essential to strengthen the capacities of both science and technologies in developing countries. There are disparities in the socioeconomic conditions of the 'developed' and 'developing' countries, which can be correlated with their scientific and technological capacities.

The developing countries need to build national institutions. For example, there are world-renowned institutions located in China and India, considered 'developing countries'. At the same time, the developing countries can come together and pool their financial and scientific resources to build international science centres. The participating countries can build and collectively operate them. Such international institutions were created in Europe. This was part of the reconstruction of science in Europe after the devastation during the Second World War.

There is cause for concern and optimism according to Egyptian-born Ahmed Zewail, now at California Institute of Technology, who was awarded the 1989 King Faisal International Prize for Science (in the subcategory physics) with Theodor Wolfgang Hänsch and the Nobel Prize in Chemistry in 1999 (unshared). Ahmed Zewail, in his article, says

> In the past five years, the scientific community worldwide has published about 3.5 million research papers. Europe's share is 37%, the USA share is 34% and the Asia-Pacific's share is 22%. Rest of the globe representing 70%–80% of the world's population living largely in developing countries has contributed less than 7% of these scientific articles [1].

Ahmed Zewail further quotes Ehsan Masood,

> Countries in the Arab World can and must do much more. They have both the money and the human resources [2].

It has been pointed out by *Science Watch*, based on the papers indexed by Thomson ISI, that the scientific output of the countries in the Middle East has been on the rise [3]. The trend is positive but far short of the output in the developed countries. The Organization of Islamic Conference (OIC) is a group of 57 countries with predominantly Muslim populations. They are home to 1.6 billion people, close to a quarter of the world population. The OIC countries' expenditure on science is about 0.2% of their *gross national product* (GNP) on *research and development* (R&D). This is much below the global average of 1.7%. Many of the OIC countries have invested good amounts in undergraduate education, but are weak at the postgraduate level. In the year 2012, the OIC countries had contributed only 2.4% to global research expenditure; 6% to academic publications; and 1.6% to patents [4]. The lack of science in the OIC countries has serious implications on the economies of these countries. Some of the poorest countries on the planet are from the OIC. Here, it is relevant to note the case of the African continent as it has about 20% of the world's area and 16% cent of the world's population. Africa is naturally blessed with many natural resources and is the principal exporter of several minerals crucial for industrial development. However, with such varied natural resources, the continent of Africa is weak and struggling with poverty.

This is a time for critical thinking and soul-searching among Arabs – in particular, for leaders of Muslim countries, who can play a critical role in building science capacities. A millennium ago, Arab sciences led the world. During the period 750–1100 CE, the Arabs were leaders in the various fields of knowledge then known. This period marked the gathering of knowledge from ancient sources in other civilisations, as well as translation of these bodies of knowledge into Arabic. These activities were carried out in bureaus of translations and centres of advanced learning (*Bait-ul-Hikmas*) under royal patronage. Thus, the Arabs made remarkable contributions to sciences and human civilisation. While Europe languished in the Dark Ages, Arab scholars had pioneering developments in algebra, medicine and the study of modern astronomy. The centres of advanced learning in the Arab lands attracted the Europeans. The Europeans took the steps of acquiring the body of knowledge preserved in Arabic and translating it into Latin and other European languages. During the period 1100–1400, the Arabs shared their scientific superiority with the emerging Europeans. Starting in the fifteenth century, the Arabs started to decline in science. Examining the causes of this decline is beyond the scope of this article. We shall just note that, paradoxically, the period of decline corresponds to the period of the great Muslim empires including the Sufvi in Iran, Osmani in Turkey and the Mughal in India. By about the sixteenth century, there was an absolute decline. The comprehensive historical accounts are to be found in the encyclopaedic works of Gibb [5] and Sarton [6]. During the *Golden Age of Science in Islam* (eighth to thirteenth centuries), the Arab scholars made the highest contributions to the sciences then known. However, today, the region is a scientific desert. In many of the Arab states, oil wealth has allowed the construction of fabulous cities, magnificent mosques, grand amusement parks and sumptuous shopping malls. However, in the same states, hardly any scientific infrastructure has emerged. The times are changing and many Arab states are now realising that the petrodollars will not keep flowing forever. Their investment in R&D is on the rise, though slowly. It is time to build institutions [4].

Along with the numerous Universities and Institutions, the developed countries have many *Learned Societies*. These actively promote the cause of science, through their publications, meetings and the system of awards and fellowships. There are yet to be enough of such academies in the Arab lands with an assured supply of funds. Islamic countries need to focus on creating centres of excellences in science and technology and other disciplines accompanied with scholarships to attract bright young minds. This will create the required human capital needed to combat illiteracy and poverty. For the countries of the 'Third World', the path to sustainable development is tortuous and obstacles formidable. Nevertheless, Europe offers examples that can be emulated, as described in the following sections.

RECONSTRUCTION OF EUROPEAN SCIENCE AFTER WORLD WAR II

The two world wars had severely affected the educational institutions across Europe and triggered a large-scale migration of scientific personnel to North America. This period coincided with the revolution in physics. The quest to understand the fundamental structure of matter was beyond the reach of small laboratories run and operated by a few individuals. There was (and still continues to be) a demand to increase the energy of the particles in the colliding beam experiments. This necessitated the construction of large particle accelerators (then circular) several kilometres in size. The building and running of such facilities requires huge funds and a large number of scientific personnel. Thus, the era of large-scale laboratories was born. In Europe, the idea was to build such facilities jointly by pooling financial and technological resources. The genesis of an international European laboratory can be traced back to the 1940s. This idea, though brilliant, took several years to mature into the gigantic enterprise, now known by the French acronym CERN: Conseil Européen pour la Recherche Nucléaire, which in English is the European Laboratory for Particle Physics. Its location at birth was in Geneva, Switzerland and now it has grown many times in size and grown across the border into France! Initially, the founding fathers of CERN faced opposition from the governments and even from the scientific community including several Nobel Laureates. The conventions to establish CERN were signed by 12 founding member countries on 1 July 1953, under the auspices of UNESCO. The conventions establishing CERN entered into force on 29 September 1954. Now, CERN has 21 member states: Austria, Belgium, Bulgaria, the Czech Republic, Denmark, Finland, France, Germany, Greece, Hungary, Israel, Italy, the Netherlands, Norway, Poland, Portugal, the Slovak Republic, Spain, Sweden, Switzerland and the United Kingdom. Creation of an international laboratory in Europe was the brainchild of Edoardo Amaldi [7,8] accompanied by the persuasion of a number of scientists and diplomats. The costs of building and running CERN are borne by the member countries according to the standard GNP formula (pay in proportion to the GNP of each country). Today, CERN is the world's largest physics research facility and has about 3,000 employees. CERN'S facilities are used by about 12,000 scientists, which accounts for half the global share of particle physicists. This diverse set of scientific personnel represents over 500 universities from 70 countries with 120 nationalities [9]. CERN has fulfilled the primary goal of advancing our understanding of the structure of matter. Moreover, it has produced remarkable technological feats. CERN's accomplishments include: the ubiquitous *World Wide Web*; *Grid Computing*, which has applications in several areas including climate studies and genome analysis; *Electronic Detectors* with potential applications in medical x-ray imaging [10].

The success of CERN sets a trend of several other joint initiatives in Europe. These include the European Space Agency (ESA) and the European Synchrotron Radiation Facility (ESRF). Synchrotrons are large circular machines, where the circulating electrons produce a special radiation, known as the synchrotron radiation (SR). SR is about a billion times brighter than the clinical x-rays and has certain other properties leading to applications not possible with the conventional x-rays. The idea of a European synchrotron facility can be traced to 1975. Its construction started in 1988, in Grenoble, France. A formal inauguration took place on 30 September 1994. ESRF is jointly supported by 21 countries. The thirteen 'Contracting Party' countries are: France, Germany, Italy, the United Kingdom, Russia, Spain, Switzerland, Belgium, the Netherlands, Denmark, Finland, Norway and Sweden. Eight countries having 'Bilateral Agreements' are: Israel, Austria, Poland, Portugal, the Czech Republic, Hungary, Slovakia and South Africa. The annual budget of about eighty million Euros is shared by all the aforementioned countries. The budget sharing for ESRF is different from that for CERN, which uses the standard GNP formula. The ESRF facilities enable research in chemistry, physics, materials and life sciences among others by over 3,000 scientists. Now, it attracts 7,000 user visits leading to over 2,000 publications per year [11].

CERN, ESRF and ESA served as a platform for the governments across Europe to work jointly. This undoubtedly had a bearing towards the formation of the European Union.

SESAME PROJECT

Let us now explore an excellent example of collaborative effort that includes several participating countries from the Middle East. It is to be noted that synchrotron light sources produce very intense pulses of light/x-rays, which allow detailed studies of objects at the atomic level with a precision that is not possible by other means such as traditional x-rays and lasers. The extraordinary power of synchrotron light has immensely impacted fields including: archaeology, biology, chemistry, environmental science, geology, medicine and physics. Establishing synchrotron facilities costs billions of dollars and requires substantial technical expertise necessitating international collaborations. There are about 75 SR facilities in various stages of operation, construction and planning in 26 countries. In 1997, Germany took the decision to replace one of its eight fully functioning synchrotrons with a more powerful facility. Germany could have sold the old synchrotron as scrap, but opted to gift it to the Middle East region. This gave birth to the Middle East synchrotron. The Middle East synchrotron is known by the apt acronym *SESAME* meaning achieving what is normally unattainable, familiar to us from the *Arabian Nights Entertainments.* Scientifically, SESAME stands for 'Synchrotron-light for Experimental Science and Applications in the Middle East'. It took several years for the SESAME project to mature. In January 2003, a groundbreaking ceremony took place in the presence of King Abdullah of Jordan and Koïchiro Matsuura, the then Director-General of UNESCO [12]. SESAME is located close to the Capital Amman. It is modelled after CERN under the auspices of UNESCO. The statutes came into effect in April 2004. The current Members (2017) of SESAME are Cyprus, Egypt, Iran, Israel, Jordan, Pakistan, Palestine and Turkey. Current Observers (2017) are Brazil, Canada, China, the European Union, France, Germany, Greece, Italy, Japan, Kuwait, Portugal, the Russian Federation, Spain, Sweden, Switzerland, the United Kingdom and the United States of America [12]. The list of countries is sure to grow. SESAME is sure to promote international scientific cooperation in the region, just as CERN had done it in Europe. The planned research programmes include: x-ray imaging, materials characterisation, archaeological microanalysis, structural molecular biology, medical applications, environmental science, micro-electromechanical devices, as well as surface and interface science. A comprehensive account of the SESAME Project is to be found in References 13 to 15. Now, the continent of Africa is the only region (other than the Antarctica!) without a synchrotron.

NEED TO CREATE REGIONAL SCIENCE CENTRES

The idea of regional science centres in the 'developing countries', particularly the Arab and Muslim countries, is simple to state but very challenging to accomplish.

In the light of the previous sections, the reader may feel that we are suggesting an ESRF or CERN. The choice of the fields at the proposed science centres shall be based on the local conditions. The same holds true for the size of the centres. It is quite possible to do world-class science in certain disciplines including mathematics, theoretical physics and mathematical biology, with minimal infrastructure. Such disciplines do not require any laboratories and can be managed with modest budgets and infrastructure. A prime example of one such institution is the celebrated Abdus Salam International Centre for Theoretical Physics (Abdus Salam ICTP) in Trieste, Italy [16].

The name of ICTP is synonymous with its founder, the 1979 Nobel Laureate Abdus Salam. Abdus Salam was born in 1926 in the small town of Jhang in the Indian subcontinent. In 1947, after the partition of the subcontinent, Jhang became part of Pakistan. Salam did his PhD in the UK and returned to Pakistan in 1951. In Pakistan, Salam found it difficult to carry out research in the absence of adequate support such as library facilities and conferences. A few years later, he returned to the UK and established a research group at the Imperial College, London. On the basis of his experience, Salam had the brainchild to create an international centre that would assist physicists working in the developing countries. The physicists would visit the centre annually and continue to work in their home countries for the rest of the year. Periodic stays at the centre would enable the visitors to keep abreast with the latest developments. This would also save the developing countries from brain drain. The brilliant idea became a reality in 1964 when ICTP was established in Trieste, Italy. ICTP was established with open-hearted support from the Italian government under the aegis of the International Atomic Energy Agency (IAEA). Starting in 1970, UNESCO also came together to support ICTP.

Over five decades, several scientific organisations have come up in and around ICTP. These include: the International School of Advanced Study (SISSA); The World Academy of Sciences (TWAS, which was founded in 1983 with the name Third World Academy of Sciences); and Third World Organization for Women in Science (TWOWS). Together, they constitute the 'Trieste Science System'. ICTP has active programmes for visitors. Besides, ICTP gives prestigious awards. Thus, ICTP is fostering science in the developing countries, which is evident in the following figures:

- ICTP has hosted over 2,000 scientific activities on its premises including introductory training schools and advanced workshops.
- ICTP has also hosted over 100,000 scientific visitors. About half of them came from developing countries and many of them regard ICTP as a scientific home away from home.
- The aforementioned visits have led to the publication of thousands of research papers.
- Statistics point to the fact that any person with a PhD in physics in the continent of Africa is invariably linked to ICTP.
- The speakers at ICTP events have included over eighty Nobel Laureates as well as recipients of other prestigious awards.

In 2015, ICTP had 5,670 visitors from 144 nations in its 51 training programmes conducted on its campus; besides, there were 21 training programmes in developing countries. During 1970–1998, Middle Eastern national have benefited by about 2,500 visits totalling to about 3,500 person months. In five decades, ICTP has progressed from a *vision to a* system. Abdus Salam died in 1996. In November 1997, in a fitting tribute, the ICTP was renamed as Abdus Salam ICTP. With its deep commitment and involvement, the ICTP has been evolving its programmes. The related examples include; a *Conference on Physics of Tsunamis* (in March 2005, within three months of the devastating Tsunami in December 2004); and the annual workshops on *Entrepreneurship for Physicists and Engineers from Developing Countries* (since 2006). In the year 2012, the ICTP budget was €27 million. With such budgets, ICTP is able to organise so many activities, offer grants, institute prizes and so forth. Such budgets will not be difficult for many of the developing countries and certainly not for the countries from the Arab lands. In the world of institutions, ICTP serves as a 'role model' and there are several centres modelled after it. These include: Asia Pacific Center for Theoretical Physics (APCTP), Pohang, Korea; and ICTP South American Institute for Fundamental Research (ICTP-SAIFR), São Paulo, Brazil. Such institutions have reduced brain drain and have also attracted back the immigrant scientists to their countries of origin. ICTP is a viable model for the developing countries. The proposed regional synchrotron facilities are another possibility [17–21].

INTERNATIONAL YEAR OF LIGHT

The International Year of Light and Light-based Technologies (IYL2015) was an enormously successful large-scale initiative leading to over 5,000 activities such as arts and science conferences, art and science projects, exhibitions, laser shows, active learning workshops and festivals [22–24]. The aforementioned activities were also conducted in schools and rural areas. These activities were attended by millions in about a hundred participating countries. The theme of light served as a unifying attractor enabling the IYL2015 to bring diverse personnel on a common platform. They could work together on the focal theme of light and address how the power of light science can be used to combat challenges to sustainable development. The potential legacies of the IYL2015 include the proposal to have an UN designated *International Day on Light and Light-based Technologies* [24]. IYL2015 was represented by only 94 National Nodes (Table 15.1), involved in organising events and outreach programmes in their countries. This figure is to be compared with the 195 Member States of UNESCO along with its 10 Associate Members.

These figures are somewhat paradoxical, with over half the countries missing the opportunity to participate in the IYL2015. Half the number of countries translates to three quarters of the population! We can conclude that about 75% of the world population of seven billion missed the IYL2015. This is to be further correlated to the figure of 1.1 billion people, who are yet to have access to electricity. This population of 1.1 billion is still using conventional sources of light. The lack of participation by the populous countries points to the inadequacies of the optics community (to which the authors belong) and the scientific community at large. The IYL2015 had the distinct advantage, as it was endorsed by numerous international scientific unions including

TABLE 15.1

National Nodes of the International Year of Light

94 Participating Countries

A	Algeria; Andorra; Argentina; Armenia; Australia; Austria
B	Bangladesh; Belgium; Bolivia; Bosnia and Herzegovina; Brazil; Bulgaria
C	Cameroon; Canada; Chile; China, Hong Kong; China, Taipei; Colombia; Costa Rica; Croatia; Cuba; Cyprus; Czech Republic
D	Denmark; Dominican Republic
E	Ecuador; Egypt; El Salvador; Estonia
F	Fiji; Finland; France
G	Germany; Ghana; Greece
H	Honduras; Hungary
I	Iceland; India; Indonesia; Iran; Iraq; Ireland; Israel; Italy
J	Japan
K	
L	Latvia; Liberia; Lithuania
M	Malaysia; Mauritius; Mexico; Mongolia; Morocco
N	Nepal; Netherlands; New Zealand; Nigeria; Norway
O	Oman
P	Pakistan; Panama; Peru; Philippines; Poland; Portugal; Puerto Rico
Q	Qatar
R	Republic of Korea; Republic of Moldova; Romania; Russia
S	Saudi Arabia; Senegal; Serbia; Singapore; Slovakia; Slovenia; South Africa; Spain; Sudan; Sweden; Switzerland
T	Thailand; Tonga; Tunisia; Turkey
U	United Arab Emirates; United Kingdom; United States of America; Uruguay
V	Venezuela; Vietnam
Y	
Z	

the International Council of Science. Moreover, it was administered in collaboration with UNESCO's 'International Basic Sciences Program', with the Global Secretariat at the Abdus Salam ICTP in Trieste, Italy. Light science is undoubtedly one of the most accessible themes to promote interdisciplinary education and industrial collaboration. Let us recall that ICTP is a UNESCO Category 1 Institute, well known for its flourishing outreach programmes. However, even this was not able to stimulate the much required participation from the developing countries in Africa, Asia, the Middle East and South America. The situation is further offset by the fact that some of the 94 enrolled countries did not take an active part. When we go through the IYL2015 calendar, some of the 94 national nodes have very few events listed under them. The scholarly societies, government and non-government organisations need to introspect on this failure. It is strange to note that the central theme of light could not operate on a much wider scale in some parts of the globe. It is time for the scientific community, in particular the founding fathers of the IYL2015, to reflect on this state of affairs. The organisers of the forthcoming international years and other vehicles of outreach programmes need to examine the various aspects of the insufficient participation [25].

CONCLUDING REMARKS

Large-scale joint scientific projects have served as vehicles of international cooperation. Such projects bring together diverse countries. This has been demonstrated in several projects and the prime example is that of CERN. The CERN laboratory brought together those countries that were at war just a few years back. CERN is perhaps the most apt example but there are other examples, including the ESRF and the ESA who contributed immensely towards the creation of the European Union. The role of scholarly societies in international collaborations is worth noting. A relevant example is that of the European Science Foundation (ESF). The ESF was established in 1974 and has promoted science across Europe. Today, ESF boasts of 72 Member Organisations from 30 European countries comprising national science bodies and research councils among others. ESF addresses the crucial issues of common concern such as the functioning of joint research facilities and science policy. ESF also has the *Standing Committee for the Social Sciences* and the *Standing Committee for the Humanities*, which assist in implementing science policies. With all the benefits and services, the annual budget of ESF is about €53 million (year 2008). It has to be borne in mind that it is difficult for many individual countries to execute large-scale science projects, thus necessitating international collaborations. This is so in view of the budgetary limitations and limited scientific resources. As discussed in the preceding sections, the developing countries and the Arab countries in particular can closely study the European institutions, in their own perspective. Then they can work towards establishing such facilities in their regions. Once established, the region will also receive benefits from the commercial and technological spin-offs [26].

The Middle East has joined the elite list of 26 countries which have synchrotrons. However, the trail to this status is based on a German donation with financial assistance from several Western countries! The SESAME facility is an excellent opportunity for regional cooperation across the Middle East, particularly as there is no other synchrotron within the region or nearby. The Iranian Light Source Facility is in the design stage [27,28]. Ideally, there needs to be an *Arab Synchrotron*, built from scratch in Arab lands with its own resources. The Arab scientists need to work together towards the scientific advancement in the region. This is sure to ensure a better future for the region and thereby to its people. On 16 May 2017, the SESAME light source was officially opened by His Majesty King Abdullah II of Jordan. The ceremony included ambassadors, ministers, former and current directors of CERN along with Irina Bokova, the director-general of UNESCO [12].

The need for creating international facilities in the African continent is very urgent. It takes years and decades to finalise the details. Hence, an early start is crucial [20,21]. The just concluded International Year of Light would have been an excellent opportunity to work towards accelerator-based light sources [27,28]. The year-long celebrations of light saw ample coverage of the pioneering contributions of Mediaeval Arab scientists to optics, particularly Ibn al-Haytham [29–35].

The picture of the scientific resources in the Arab and Muslim countries is dismal but they can definitely plan and fill the gap. Europe and other nations received results by giving due importance to science and technology. The Holy Quran in 750 of its verses (i.e., over one-eighth of its total verses) urges the believers to reflect on and study nature deeply. There is all along an emphasis on reason. Science enables us to understand nature and its divine design. Science leads to material benefits. Scientific centres would be an excellent vehicle of international cooperation across the Muslim countries. Strength in science constitutes a measure for the honourable survival of a nation. It is time for each nation to strengthen its scientific base. This requires scientific institutions and universities with research programmes. Half the manpower of any nation needs to be grounded with scientific training. The path to this status requires generous government patronage and adequate allotment following the norms practised by the 'developed countries'. The required norms are: 2%–3% of the GNP on R&D; 5% of GNP on basic education; the defence related research is in addition to these. Besides, they have to spend 5% of the GNP on health. As of now, half the OIC countries are meeting the norms on education in terms of expenditure. When it comes to health and R&D, the OIC countries are much below the norms. Table 15.2 has the figures for civil research for different regions. The Middle Eastern countries have established several prestigious prizes for sciences including the *King Faisal International Prize* [36]; *UNESCO Sultan Qaboos Prize for Environmental Preservation* [37]; and the *Mustafa Prize*, which was launched by Iran in December 2015 [38]. They can similarly create scientific institutions. We need to strive to create a *Commonwealth of Sciences for Islamic Countries* [39,40]. Without these, the Muslim countries and their citizens will never be able to lead a normal existence, full of dignity in the comity of nations. One of the many outcomes of the International Year of Light was the formation of an *International Working Group 'Ibn al Haytham'* (IWG). A conference was held at the UNESCO Headquarters, Paris, France under the auspices of IWG [34,41]. Now, IWG has grown into the *Ibn al Haytham LHiSA International Society*, where LHiSA stands for Light: History,

---

TABLE 15.2
Statistical Data for Regions 2014

Region	Researchers (Per Million Inhabitants)	Publications Per Million Inhabitants	Expenditure on R&D (% of GNP)
Africa	168	29	0.45
Arab States	417	82	0.30
Asia	786	118	1.62
Europe	2,942	609	1.75
North America	4,034	1,013	2.71
Oceania	3,219	1,389	2.07
Latin America	488	112	0.69
World	1,083	176	1.70

Source: UNESCO Science Report, Towards 2030, UNESCO Institute of Statistics (2016).

Science and Applications [42]. IWG and LHiSA are the brainchild of the Optical scientist Prof. Azzedine Boudrioua from the University of Paris. LHiSA is an excellent platform to carry out the ideas of creating international science centres not only in Arab and Muslim countries but across the developing countries. Almighty Allah has promised, *He does not let the efforts of those who strive, go waste* [43].

ACKNOWLEDGEMENTS

We would like to thank profusely Prof. Boudrioua Azzedine for inviting us to the Conference. We are very thankful to the UNESCO for providing us with complete financial support towards air travel and local hospitality. In the absence of such support, our participation would have been difficult. We appreciate the warm hospitality which we enjoyed throughout our stay for the *Landmark September Event.*

REFERENCES

1. Ahmed Zewail, Global science and global peace, *Europhysics News*, 35 (1), 2004. http://dx.doi.org/10.1051/epn:2004104
2. Ehsan Massod, Arab science: Blooms in the desert, *Nature*, 416, 120–122, 2002. http://dx.doi.org/10.1038/416120a
3. Middle East, *Global Research Report*, Science Watch, 2011. http://sciencewatch.com/grr/middle-east (accessed date: 17/05/2017).
4. Nidhal Guessoum and Athar Osama, Institutions: Revive universities of the Muslim world, *Nature*, 526 (7575), 634–636, 2015. http://dx.doi.org/10.1038/526634a
5. Hamilton Alexander Rosskeen Gibb, *The Encyclopaedia of Islam* (Brill Academic Publishers, Leiden, The Netherlands), 1986.
6. George Sarton, *Introduction to the History of Science*, in four volumes (Williams & Wilkins, Baltimore), 1962.
7. Herwig Schopper and Sameen Ahmed Khan, CERN's early history revisited, *Physics Today*, 58 (4), 87–89, 2005. http://dx.doi.org/10.1063/1.4796963
8. Carlo Rubbia, *Edoardo Amaldi Scientific Statesman*, CERN Report, CERN-91-09, Geneva, 1991 (European Organization for Nuclear Research); http://cds.cern.ch/record/228364/files/CERN-91-09.pdf (accessed date: 17/05/2017).
9. CERN Website: http://www.cern.ch/ (accessed date: 17/05/2017).
10. Sameen Ahmed Khan, CERN and the birth of World Wide Web, *Youth Observer*, Supplement to Oman Observer, 24 (174), 6, 2005.
11. ESRF Website: http://www.esrf.fr/ (accessed date: 17/05/2017).
12. SESAME Website: http://www.sesame.org.jo/ (accessed date: 17/05/2017).
13. Sameen Ahmed Khan, The Middle East Synchrotron Laboratory and India, *Current Science*, 80 (2), 130–132, 2001. http://www.iisc.ernet.in/currsci/jan252001/130.pdf (accessed date: 17/05/2017).
14. Azher Majid Siddiqui and Sameen Ahmed Khan, SESAME, the first international science centre in the Middle East: A step towards the Renaissance of science in the Islamic countries, *Journal of Islamic Science*, 17 (1–2), 9–34, 2001 (Muslim Association for the Advancement of Science, Aligarh, India).
15. Sameen Ahmed Khan, The Middle East Synchrotron Facility can bring regional cooperation, *Digest of Middle East Studies (DOMES)*, 11 (2), 57–71, 2002. http://dx.doi.org/10.1111/j.1949-3606.2002.tb00457.x
16. ICTP Website: http://www.ictp.it/ (accessed date: 17/05/2017).

17. Sameen Ahmed Khan, Need to create Regional Synchrotron Radiation Facilities (RSRF), *IRPS Bulletin*, 17 (2), 7–13, 2003 (International Radiation Physics Society). http://www.canberra.edu.au/irps/bulletin/docs/Vol-28-No-1-April,-2014.pdf (accessed date: 17/05/2017).

18. Sameen Ahmed Khan, Prospects for an Asian Accelerator Laboratory, *AAPPS Bulletin*, 12 (2), 21–27, 2002 (Association of Asia Pacific Physical Societies). http://aappsbulletin.org/ (accessed date: 17/05/2017).

19. Sameen Ahmed Khan, When will there be an Asian Accelerator Laboratory? *ICFA Beam Dynamics Newsletter*, 28, 49–54, 2002 (International Committee for Future Accelerators). http://icfa-usa.jlab.org/archive/newsletter/icfa_bd_nl_28.pdf (accessed date: 17/05/2017).

20. Sameen Ahmed Khan, The Middle East Synchrotron is launched; Armenian Synchrotron; To launch the African Synchrotron Programme, *AAPPS Bulletin*, 13 (2), 35–36, 2003 (Association of Asia Pacific Physical Societies). http://aappsbulletin.org/ (accessed date: 17/05/2017).

21. Sameen Ahmed Khan, Ground breaking for the Middle East Synchrotron; Armenian Synchrotron; Time to launch the African Synchrotron Research Programme, *ICFA Beam Dynamics Newsletter*, 30, 88–89, 2003 (International Committee for Future Accelerators). http://icfa-usa.jlab.org/archive/newsletter/icfa_bd_nl_30.pdf (accessed date: 17/05/2017).

22. Sameen Ahmed Khan, 2015 declared the International Year of Light and Light-based Technologies, *Current Science*, 106 (4), 501, 2014. http://www.currentscience.ac.in/Volumes/106/04/0501.pdf (accessed date: 17/05/2017).

23. Sameen Ahmed Khan, *International Year of Light and Light-based Technologies* (LAP LAMBERT Academic Publishing, Germany), 96 pp., 2015. http://www.lap-publishing.com/ and http://isbn.nu/9783659764820/ (accessed date: 17/05/2017).

24. Jorge Rivero González and John Dudley, Inspired by light: Close of the International Year of Light, *Europhysics News*, 47 (2), 6–7, 2016. http://www.europhysicsnews.org/articles/epn/pdf/2016/02/epn2016-47-2.pdf (accessed date: 17/05/2017).

25. Sameen Ahmed Khan, Reflecting on the International year of light and light-based technologies, *Current Science*, 111 (4), 627–631, 2016. http://dx.doi.org/10.18520/cs/v111/i4/627-631

26. Sameen Ahmed Khan, 2004 – the year of jubilees; fifty years of CERN; forty years of ICTP, ten years of ESRF User Operation, *ICFA Beam Dynamics Newsletter*, 37, 12–18, 2005 (International Committee for Future Accelerators). http://icfa-usa.jlab.org/archive/newsletter/icfa_bd_nl_37.pdf (accessed date: 17/05/2017).

27. Sameen Ahmed Khan, X-rays to synchrotrons and the International Year of Light, *IRPS Bulletin*, 28 (1), 9–13, 2014 (International Radiation Physics Society). http://www.canberra.edu.au/irps and http://radiationphysics.org/ (accessed date: 17/05/2017).

28. Sameen Ahmed Khan, Particle accelerators and the International Year of Light, *ICFA Beam Dynamics Newsletter*, 63, 9–15, 2014 (International Committee for Future Accelerators). http://icfa-usa.jlab.org/archive/newsletter/icfa_bd_nl_63.pdf (accessed date: 17/05/2017).

29. Alistair Kwan, John Dudley and Eric Lantz, Who really discovered Snell's Law? *Physics World*, 15 (4), 64, 2002. http://physicsworldarchive.iop.org/full/pwa-pdf/15/4/phwv15i4a44.pdf and http://web.physik.uni-rostock.de/optik/de/WhoDiscoveredSnellsLaw.pdf (accessed date: 17/05/2017).

30. Roshdi Rashed, A Pioneer in Anaclastics: Ibn Sahl on Burning Mirrors and Lenses, *ISIS*, 81 (3), 464–491, 1990. http://dx.doi.org/10.1086/355456

31. Roshdi Rashed, A polymath in the 10th century, *Science*, 297, 773, 2002. http://dx.doi.org/10.1126/science.1074591

32. Sameen Ahmed Khan, Medieval Islamic achievements in optics, *Il Nuovo Saggiatore*, 31 (1–2), 36–45, 2015 (Società Italiana di Fisica, the Italian Physical Society). http://prometeo.sif.it/papers/online/sag/031/01-02/pdf/06-percorsi.pdf (accessed date: 17/05/2017).

33. Sameen Ahmed Khan, Medieval Arab contributions to optics, *Digest of Middle East Studies (DOMES)*, 25 (1), 19–35, 2016. http://dx.doi.org/10.1111/dome.12065

34. Sameen Ahmed Khan, The International Year of Light and Light-based Technologies (*Report of the Conference, The Islamic Golden Age of Science for today's Knowledge-based Society: The Ibn Al-Haytham Example*, UNESCO Headquarters, Paris, France, 14–15 September 2015), *American Journal of Islamic Social Sciences (AJISS)*, 33 (1), 160–163, 2016.

35. Sameen Ahmed Khan, Medieval Arab achievements in optics, Chapter-14 in *Light-Based Science: Technology and Sustainable Development, Proceedings of The Islamic Golden Age of Science for Today's Knowledge-Based Society: The Ibn Al-Haytham Example* (14–15 September 2015, UNESCO Headquarters, Paris, France), Editors: Azzedine Boudrioua, Roshdi Rashed and Vasudevan Lakshminarayanan, pp. 195–204 (CRC Press, Taylor & Francis, UK), 2017. http://isbn.nu/9781498779388/ and https://www.crcpress.com/Light-Based-Science-Technology-and-Sustainable-Development/Lakshminarayanan-Roshdi-Azzedine/9781498779388

36. Sameen Ahmed Khan, 2017 King Faisal International Prize for Science and Medicine, Current Science, 112 (6), 1088–1090, 2017; ibid 110 (7), 1140–1141, 2016; ibid 108 (7), 1202–1203, 2015; ibid 106 (4), 500, 2014; ibid 104 (5), 575, 2013. http://www.current-science.ac.in/Volumes/112/06/1088.pdf (accessed date: 17/05/2017).

37. Sameen Ahmed Khan, 2015 UNESCO Sultan Qaboos Prize for Environmental Preservation, *Current Science*, 110 (1), 15, 2016. http://www.currentscience.ac.in/Volumes/110/01/0015.pdf (accessed date: 17/05/2017).

38. Sameen Ahmed Khan, Iran launches the Mustafa Prize for Sciences, *Current Science*, 110 (6), 961, 2016. http://www.currentscience.ac.in/Volumes/110/06/0961.pdf (accessed date: 17/05/2017).

39. Abdus Salam, Scientific Thinking: Between Secularisation and the Transcendent, in *Ideals and Realities*, third edition, pp. 283. Editors: C H Lai and A Kidwai (World Scientific, Singapore), 1989.

40. Abdus Salam, The Gulf University and the Science in the Arab-Islamic Commonwealth, in *Renaissance of Sciences in Islamic Countries*, pp. 31, Editors: H R Dalafi and M H A Hassan (World Scientific, Singapore), 1994.

41. Sameen Ahmed Khan, International Year of Light and History of Optics, *Chapter-1 in Advances in Photonics Engineering, Nanophotonics and Biophotonics*, Editor: Tanya Scott (Nova Science Publishers, New York), 2016, pp. 1–56, 2016. http://isbn.nu/9781634844987 and https://www.novapublishers.com/ (accessed date: 17/05/2017).

42. Ibn al-Haytham LHiSA International Society. https://www.ibnalhaytham-lhisa.com/.

43. Sameen Ahmed Khan, Need to create science centres in the Muslim world, *Young Muslim Digest*, 26 (7), 13–16, 2004.

16 SESAME
The First Synchrotron Light Source in the Middle East and Neighbouring Regions

Gihan Kamel

CONTENTS

Introduction 221
Ibn al-Haytham and His Followers: A Non-Stop Source of Inspiration for
SESAME People 222
Synchrotron Light Sources 224
Timeline of SESAME 224
International Support and Training Opportunities 225
UNESCO Support 226
Installation of SESAME's Storage Ring 227
References 228

Long is the way and hard, that out of Hell leads up to light.

John Milton

INTRODUCTION

Turning the wheel of history, seeking the initial steps towards the understanding of the optical systems based on the invaluable contribution of the Muslim and Arab Scholar, Ibn al-Haytham, enthusiastic scientists succeeded in developing modern eyes capable of resolving the matter down to its atoms, known as synchrotron light sources. They are super microscopes with different experimental techniques known as beamlines, which are powerful enough to reveal the finest details about the unknowns of physics, chemistry, biology, pharmacy and biomedicine, as well as materials science, environment, art restoration, cultural heritage and some other arenas. Synchrotron light for Experimental Science and Applications in the Middle East (SESAME) is one of those eyes, and is the first facility of its kind in the Middle East constructed in Jordan.

Following the significant model of CERN, the European Organization for Nuclear Research, SESAME (Figure 16.1), started a long journey since 1997. The current members are Cyprus, Egypt, Iran, Israel, Jordan, Pakistan, the Palestinian Authority

221

FIGURE 16.1   SESAME building, 35 km northwest of Amman, Allan, Jordan. (Image courtesy of SESAME.)

and Turkey, whereas Brazil, Canada, China, the European Union, France, Germany, Greece, Italy, Japan, Kuwait, Portugal, the Russian Federation, Spain, Sweden, Switzerland, the United Kingdom and the United States are the observers. It is established that this third generation light source seeks Excellence of Science, along with bridging of the gaps between its culturally diverse and politically conflicting societies. Various opportunities to SESAME objectives are witnessed; out of them is the brain-drain problem, extremely observed in the region.

Undoubtedly, it had to be something incredible that stimulated the SESAME members (Figure 16.2) to complement their efforts with an altered strategy using the common language of science that they all recognise and trust without any peculiar opinions. With tense diplomatic relations between countries, mutual research efforts can open new doors when others are being closed; therefore, the prerequisite for science diplomacy as a fresh vision nowadays is significantly emerging, taking many projections, and it is well acknowledged that scientists in their own way are able to create a more peaceful, prosperous world. In case of SESAME; difficulties are expected, many have been already faced, and new challenges are still there, but soon, with its planned commissioning in late 2016, the dream of hundreds will become a reality paving the way to a strong and competent community that is able to deal with its scientific challenges and hopefully beyond. Confirming the President of the SESAME Council, Professor Sir Chris Llewellyn Smith: 'It is a remarkable tribute to the spirit of cooperation in pursuit of a common goal which underwrites the project that SESAME is progressing so well during a time of external turbulence'.

IBN AL-HAYTHAM AND HIS FOLLOWERS: A NON-STOP SOURCE OF INSPIRATION FOR SESAME PEOPLE

As recognised by the recent discoveries, Ibn al-Haytham had a pronounced and inspiring influence on the world of science across generations and untill this day. In the thirteenth century, and particularly in Toledo, the Spanish municipality located

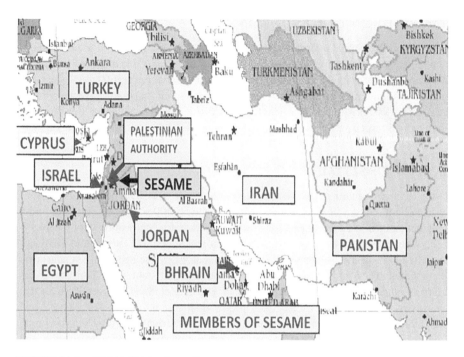

FIGURE 16.2   SESAME Members. (Image courtesy of SESAME.)

in central Spain, a historical coexistence of Christian, Jewish and Muslim intellectuals who worked compassionately to translate Arabic and Hebrew books, as well as other scientific works, into Latin. Could this inspiring plot be of any closeness to the core concept of SESAME? The answer is yes. Only by joint contributions, could the accumulated old scientific knowledge have that obvious impact on our present and future lives. Similarly, the SESAME people in all their differences are working together to achieve a common goal that is translating science into peace.

> It is well demonstrated that in today's information age, distances and languages are no longer barriers to the transfer of knowledge. With the power of technology and advancements in science, collaboration across cultures could bring massive leaps and rapid advances towards solving shared global challenges and achieving mutual benefits, while also contributing to intercultural dialogue.

Hanan Dowidar and Ahmed Salim[1]

> Science for peace only works if the science is truly excellent. My job as president of the council is to ensure that SESAME will be a first-class scientific instrument. If it is, SESAME will attract excellent scientific users from all the member countries and encourage scientists to remain in, or return to, their home region to pursue their research. These scientists will work together to produce first-class science while building personal links that cross political, cultural and religious boundaries.

Professor Sir Chris Llewellyn Smith

SYNCHROTRON LIGHT SOURCES

Light sources are capital intensive and very expensive projects requiring clear-cut technical capacities. In any synchrotron light source, the produced synchrotron radiation does cover a wide range of the electromagnetic spectrum from the infrared to the x-ray regions. This radiation is then collected out of the source machine by very sophisticated optical systems – known as beamlines – those are used to focus the light on to the experimental targets, allowing several experiments to be performed simultaneously.

TIMELINE OF SESAME

Owing to their wide impact and benefits, there are now more than 60 operational synchrotron light sources in 19 countries serving some 30,000 scientists. More are under construction or in various stages of planning. Additionally, many of the rapidly emerging economies, such as Brazil, the Republic of Korea, Singapore, Taiwan and Thailand, have already built their own synchrotron light sources.[2] The need for a synchrotron light source in the Middle East was recognised by the Pakistani Nobel laureate Abdus Salam – founder of the International Centre of Theoretical Physics – over 30 years ago, as it was also considered by CERN (European Organization for Nuclear Research) and the Middle East-based Scientific Cooperation (MESC) group headed by Sergio Fubini, a theoretician at CERN, with efforts to promote regional cooperation in science, and that also encompass solidarity and peace, started in 1995.[3]

In 1997, Herman Winick from the Stanford Linear Accelerator Center (SLAC) National Accelerator Laboratory, United States, and the late Gustav-Adolf Voss from the Deutsches Elektronen-Synchrotron (DESY), Germany, both suggested building a light source in the Middle East using the components of the decommissioned BESSY I facility in Berlin. Winick and Voss believed that BESSY I could serve a step for advanced science in a region not up to standard for scientific development, and the Middle East was their candidate.

> Our top laboratories operate under budget strictures, yet we usually have the resources to obtain the newest and best equipment for upgrades to our facilities. Isn't it better to use that old equipment to recycle the items, to put them in the hands of projects struggling for funds instead of just putting them in the corner?
>
> **Winick[4]**

The perception of science for peace in the Middle East had been running in many minds at that time, only waiting for an action to be arranged. In 1995, a meeting hosted near the Red Sea resort of Dahab by the Egyptian government, Venice Gouda, Minister of Higher Education of Egypt and the Israeli physicist Eliezer Rabinovici took an official stand of cooperation. In attendance were scientists from throughout the Middle East (Egypt, Israel, Jordan and the Palestinian Authority), and from Europe and the United States.

The destination made itself clear again when Voss encountered several Middle Eastern scientists in meetings organized in Torino by Sergio Fubini. Then, in 1998 for the MESC meeting in Uppsala, Sweden, when Winick and Voss presented the proposal for recycling BESSY I as a basis for a synchrotron light source in the Middle East. The Palestinian scientist, Said Assaf, Director-General of the Palestinian Authority's Arafat National Scientific Centre for Applied Research, West Bank was among the others who also pushed for the project. Joint efforts involved the CERN Director-General at that time, Herwig Schopper, who also worked to confirm the agreement with the German government.[4] The proposal was brought to the attention of Federico Mayor, the then director general of UNESCO (United Nations Educational, Scientific and Cultural Organization), who called a meeting in June 1999 of delegates from the Middle East and other regions. The meeting launching the project had led to the formation of the Interim Council and four Committees (Scientific, Technical, Training and Financial) to follow up this idea and prepare for the next steps. On the basis of the proposal of Said Assaf (Palestinian Authority), the acronym 'SESAME' was selected not only to evoke memories of the culture of the region, that is, 'Open SESAME', but also because it describes the Centre's Mission which is a precise description of the facility, namely to be open to all scientists of the world (SESAME).[3]

Efforts were – and are still – tremendous and scientists actively pushed for SESAME and in due course will win. In 2002, a decision has been taken to build a completely new main storage ring, with a circumference of 133.2 metres and higher energy (2.5 GeV) than would have been provided by upgrading the main BESSY I ring (1 GeV), while retaining the main elements of the BESSY I microtron, which provides the first stage of acceleration, and the booster synchrotron. The SESAME new ring has been upgraded to ensure making SESAME a third-generation machine. The groundbreaking ceremony was held in 2003, and SESAME formally came into existence in April 2004, after UNESCO had received the required instrument of acceptance of the draft statutes from six initial Members. The interim Council was soon replaced by the permanent Council followed by the approval of the rules and regulations.

SESAME's main mission is laid down in the Statutes that established the Centre that: 'SESAME shall provide for collaboration in the Middle East and the Mediterranean Region with free access to all scientists of SESAME members in relevant areas of research, being also open to scientists from the whole world, in basic and applied research using synchrotron radiation or closely related topics'. A Statement of particular interest: 'Special support will be given to work of relevance to the region and scientific and industrial research. SESAME shall not undertake classified work for military purposes or other secret research, and the results of its experimental and theoretical activities shall be ultimately published or otherwise made generally available'.[3]

SESAME is owned by its Members which have full control over its strategic development and exploitation, and major financial matters. Day-to-day management is the responsibility of the Director and Directorate, appointed by, and answerable to, the SESAME Council, on which all the Members are represented.

INTERNATIONAL SUPPORT AND TRAINING OPPORTUNITIES

With its birth, a wide international interest in and huge support for SESAME was recognised. Besides the worldwide Observers, the SESAME Advisory Committees (Scientific, Technical and Training) involve senior scientists and experts from

Canada, France, Italy, Japan, Kuwait, Spain, Switzerland, the United Kingdom and the United States, as well as ICTP. Equipment has been donated by France, Italy, Switzerland, the United Kingdom and the United States. The United Kingdom has similarly donated the essential components of five beamlines from its Daresbury synchrotron. European, American and Japanese centres are also contributing valuable assistance and advice.

A significant training programme has been more or less entirely funded by the generosity of scientific bodies that are inspired by the vision that underlies the project – such as the International Atomic Energy Agency (IAEA), ICTP, the European Synchrotron Radiation Facility (ESRF), the European Union LinkSCEEM and VI-SEEM projects, synchrotron laboratories all over the world, and some national organisations as well. The vigorous training programme is also funded by scientific bodies (the American Physical Society, the American Chemical Society, Deutsche Physikaliche Gesellschaft, the European Physical Society, the Institute of Physics and the International Union of Pure and Applied Physics) and by two foundations (the Canon Foundation for Scientific Research and the Richard Lounsbery Foundation).[2]

UNESCO SUPPORT

UNESCO is the depository of the Statutes of SESAME. In 2002, the UNESCO General Conference and Executive Board approved SESAME under the auspices of UNESCO as an independent laboratory. Within the Executive Board, SESAME received enthusiastic and unanimous support and it was expressed that it could be a model project for other regions. The Executive Board called SESAME 'a quintessential UNESCO project combining capacity building with vital peace building through science'.[3]

On 21 May 2012, Irina Bokova, Director-General of UNESCO, visited SESAME for the first time. The visit was organised to mark the formal statements signed in March 2012 by Iran, Israel, Jordan and Turkey, committing to make voluntary contributions of US$5 million each towards the construction of SESAME over the four years 2012–2015, a milestone in the project that puts SESAME well on track to the commissioning stage. 'SESAME is the embodiment of UNESCO's commitment', said the Director-General, pointing out that the project is 'about weaving a fabric of intellectual and moral solidarity across the world'. At her visit, the Director-General of UNESCO and the President of the SESAME Council signed a joint Communiqué to governments, international and national scientific organisations, scientific institutions and centres of excellence, all stakeholders in science, and the public highlighting the role of science in fostering solidarity and in building a secure and prosperous future for the region. The Directors-General of UNESCO have always played a central role in each of the milestones of SESAME along its journey. It was subsequent to his successor, Koïchiro Matsuura, having informed the Federal Minister of Education and Research of Germany in January 2000 that he was ready to take the necessary steps for the setting up of SESAME as a centre under the auspices of UNESCO, and his assurance that financing of the dismantling of the BESSY I machine in Berlin would be covered from international sources that the German Authorities gave their formal agreement to donating the BESSY I machine to SESAME. The ground-breaking ceremony of SESAME (January 2003) was held under the auspices of HM King Abdullah II of Jordan and in the presence of Koïchiro Matsuura, Director-General of UNESCO. The 'soft' inauguration (November 2008) was held under the auspices of HM King Abdullah II of Jordan and with the participation of HRH Prince Ghazi bin Mohammad and Koïchiro Matsuura.[2]

INSTALLATION OF SESAME'S STORAGE RING

During May 2013, an agreement was signed between the European Commission and CERN to supply magnets for SESAME's main ring and its powering scheme. Collaborative efforts provided by Spain, Cyprus, Israel, Pakistan and Turkey for the magnets production and testing together with the essential power supplies. In addition to that, Iran, Pakistan and Turkey have provided CERN with personnel in the form of in-kind support. The team responsible for this important milestone consisted of scientists and technicians from the SESAME region, with help from CERN as a part of the CERN-EC Support for SESAME Magnets (CESSAMag) team. As depicted in Figure 16.3, the first of the 16 cells of SESAME's ring was installed in the shielding tunnel in the Center's experimental hall delivering a historic moment for both SESAME and the Middle East, on 10 February 2016. Each cell consists of a group of magnets (dipole, quadrupoles and sextupoles) and the vacuum chamber, supported by a girder.[3]

Commissioning of the SESAME machine is expected to start during the end of 2016, and the laboratory is expected to become operational with two 'day-one' beamlines in 2017. These beamlines are the XAFS/XRF (x-ray absorption fine structure/x-ray fluorescence) beamline and the IR (infrared) microspectroscopy beamline. This success will open the path for the final goal, which is to make SESAME the first operational synchrotron light source in the Middle East and to confirm its position as a truly international research centre. In the words of an endorsement of SESAME

FIGURE 16.3 Installation of the SESAME storage ring has started in February 2016. (Image courtesy of SESAME.)

issued by 45 Nobel laureates, SESAME is expected to become 'a beacon, demonstrating how shared scientific initiatives can help light the way towards peace'.[2,3]

A Synchrotron beam was circulated for the first time on 12 January 2017 in the main ring of the pioneering SESAME. Technical efforts were combined following this first single turn, to achieve multi-turns, to store and then accelerate the beam. This is a significant milestone on the way to the Synchrotron-based research in the region. On 16 May 2017, SESAME was officially inaugurated by His Majesty King Abdullah II, in a historic moment for SESAME opening its doors to the scientists of Middle East and neighboring regions.

REFERENCES

1. Hanan Dowidar and Ahmed Salim, Celebrating optics pioneer Ibn al-Haytham – and promoting intercultural collaboration for the International Year of Light. *Elsevier Connect*, 2015.
2. Chris Llewellyn Smith, SESAME for science and peace. *Nature Photonics*, 9, 550–552, 2015.
3. Historical Highlights. Foundation of a synchrotron light source in the Middle East. SESAME official site, copyright ® SESAME.
4. Mike Perricone, SESAME: Can a recycled synchrotron become an oasis of peace in the Middle East? *Symmetry*, 1(2), 18–23, 2005.

17 Scientific Translation
A Tool from Shadow to Light

Abdulaziz Alswailem

CONTENTS

Introduction...229
Translation Role in Civilisations...230
Schools of Knowledge and Translation 230
Lessons Learnt 231
Shifting Economy to Knowledge: KSA's National Strategy 231
Scientific Awareness Outreach Programme: Science for All.......................235
    Focus Groups 235
    Collaborating Institutions...236
    Outreach Projects 236
        Acquisition of Printed Materials 236
        Multimedia 237
        Future Scientist Project 237
        Messages (20 min)...238
KACST Scientific Translation Efforts 238
Science, Technology and Innovation Key Performance Indicator (KPI)..............239
KACST Light Highlights..240
Recommendations..242
References 242

INTRODUCTION

Communication in either verbal or non-verbal is one of the effective tool in knowledge transfer between civilizations. Verbal communication uses different languages which are unique for each language [1]. It is a reality that all textual writings from literature to science involve different languages and, for some countries, cryptology. Early writings in science come in different coinage, different vocabularies and symbols. It started in early ancient civilizations and continue to the modern era. Greek and Latin were translated into different languages, including the Arabic language.

Translation is an operation performed on a language. It is a skill, a craft and a science. Translation intends to broaden and enrich knowledge by giving the same thought, idea objective, information, culture and interpretation as the original [2]. Translation in various areas of knowledge is a product of a broad research. However, translation of scientific content has a less proportion than the others such as history and novels. Scientific translation brings knowledge and information in the global spectrum of the changing science and technology.

TRANSLATION ROLE IN CIVILISATIONS

Through translation, the product (output) of one civilization is the seed (input) of a new one. Translation is a relay of human progress, development and prosperity. An example is the contribution of the Arabic language to mathematics and economics. The Arabic language is considered to be one of the ancient languages. It introduced the concept of zero (sifr in Arabic) in arithmetic from which emanated the concept of 'cipher' in all European languages to designate the zero value. This concept of zero made more concrete scientific calculations in trade, economy and in science and technology.

Historically, the time between 700 and 1300 AD considered to be the Islamic golden era, where it flourished culturally, economically and scientifically. The golden era of Arabic translation historically started in about 661–750 AD and reached its peak of translation from other languages in 700 AD to 1300 AD. The well-known House of Wisdom, founded in Baghdad, was the intellectual centre founded during the time of Caliph Harun al-Rashid (786–809 AD) as shown in Figure 17.1 that formalised the translation from other languages to Arabic. University scholars with different cultures collected books with classical knowledge written mostly in Greek and other eastern languages and translated them into Arabic. Some of the well-known translations were done by Ibn Al-haytham who is known in the west as Alhazen (965–1040). His famous translation was of the work of the Greek physician by the name of Galen. His translation of Galen's work was then translated into other languages.

SCHOOLS OF KNOWLEDGE AND TRANSLATION

Knowledge is dynamic and never ceases because of man's unending quest for learning and exploration. The School of Edessa, now Sanliurfa in Turkey, was founded in 2 AD. The school is of great importance because it is the birthplace of the Syriac literature and philosophy. The Syriac language came from the Aramaic alphabet through the Palmyrene alphabet, and has similarities with the Phoenician, Hebrew, Arabic and traditional Mongolian alphabets. The scientific ancient wisdom of the Greeks in arts, music, logic, mathematics, geometry, medicine and other literary works of Plato and Aristotle rose from the School of Edessa [3] in about 400 AD and the school was closed in 480 AD. The new centre of the ancient wisdom, the Persian Academy of Jundisabur (also known as Gundeshapur) [4], was founded and

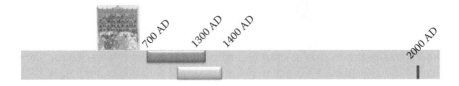

FIGURE 17.1 Foundation of the House of Wisdom and the years of important events in the birth of translation.

was the intellectual centre of the Sasanian empire. The advancement of arts, sciences and medicine was dominated by the Arabic culture. The School of Nisibis existed from the fifth to the seventh centuries absorbing the students from the School of Edessa after it had been dissolved. It gave life to the Syrian intellectual wisdom. From the ninth to the thirteenth centuries came the House of Wisdom in Baghdad and during the twelfth and the thirteenth centuries, the Toledo School in Toledo, Spain was founded. With the School of Wisdom [5] and the Toledo School, the science translation in different languages was in a rapid growth phase with the main translation work done in Toledo School. The development of the new scientific intellectual world broadened and revolutionised scientific knowledge and translation. It was a continuous process of translation, explication and commentary, criticism and correction leading to innovation.

LESSONS LEARNT

The historical development of the translation from Arabic to English and vice versa showed that there exists a knowledge gap in the process of translation. The very hindrance is core of knowledge on the language where there are problems and deficiencies in vocabularies using mediaeval and modern forms of words, symbols and for some alphabets and cryptology [6]. Scientific translation poses issues on terminologies and there is a need to choose appropriate terminology. Documentation and dissemination of terminological issues are important in addressing these issues.

History dictates that there is no language, religion, gender or borders for knowledge, science and innovation. Most of the major translation efforts throughout history were started by leaders. History shows that translation is a long-term process but the real output will be shown in the next generations.

Today's modern world on information technology and computerisation, databases and software developments are very efficient tools for harnessing instant translation with the aid of different search engines in the internet. The use of these tools and other interactive platforms can make scientific translation readily available, understandable and easy to comprehend.

SHIFTING ECONOMY TO KNOWLEDGE: KSA'S NATIONAL STRATEGY

The abundance of natural resources is balanced by the knowledge on its proper use which plays a vital role in economy. The growth of the national economy is the product of the significant contribution of innovation, technology transfer and development. The Atlas of Economic Complexity attempts to measure the amount of productive knowledge that each country holds, by visualising the differences between national economies (https://en.wikipedia.org) and illustrates through a product space shown in Figure 17.2. Using this visualisation, the map shows that in 1980, the primary and high ranking product of Saudi Arabia was crude oil as illustrated in Figure 17.3. In 2012, the production increased with the slow growth of the economy compared to South Korea, Germany and the United States in the same year as shown in Figure 17.4.

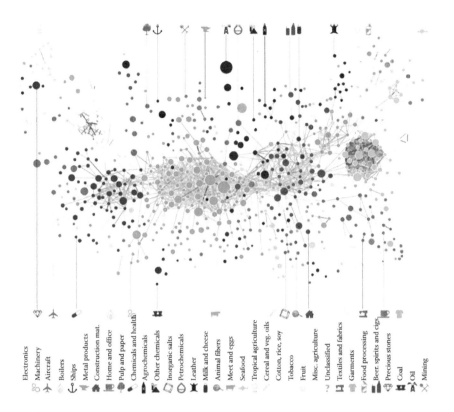

FIGURE 17.2 Pictorial representation of the product space shown in the Atlas of Economic Complexity. (Source: www.atlas.med.mit.edu)

KACST is responsible of formulating R&D policy in the Kingdom of Saudi Arabia, and developing the national science, technology and innovation strategic plan and supervising its implementation. KACST's vision for Saudi Arabia is building a knowledge-based society and economy and a globally competitive national system for innovation. It will make Saudi Arabia a world-class organisation in science and technology, fostering innovation.

The Science and Technology National Policy (STNP) is a directional framework created for the future of STI in the Kingdom. STNP comprises principles and bases from which it endeavours to achieve the long-term science and technology objectives and goals of the Kingdom of Saudi Arabia. 'The Science and Technology National Policy for Science and Technology' prepared by KACST in collaboration with the Ministry of Economy and Planning was approved by the Council of Ministers in the year 2012 G (1432 AH). The strategic direction is for the Kingdom of Saudi Arabia to reach the ranks of developed countries in STI by the year 2030 as pictorially illustrated in Figure 17.5, thus improving productivity with the long-term plan of

- Development of STI infrastructure in 2014 G (1435)
- Leadership in the field of STI in the Middle East in 2019 G (1440 AH)

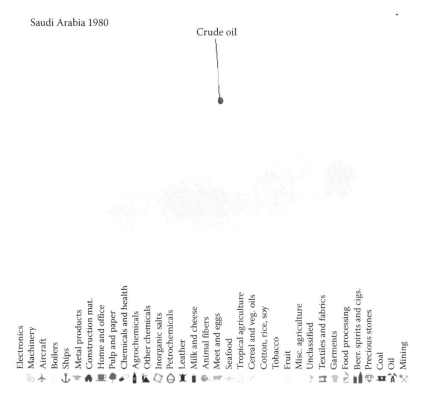

FIGURE 17.3 Product of Saudi Arabia for the year 1980 showing only crude oil as the primary product.

- Access to the ranks of industrialized countries in the field of science, technology and innovation (which means to be recognized as a player along with industrialized countries).

The National Science Technology and Innovation Plan (NSTIP) was created as a product of STNP. The goals of the national plan for science technology and innovation are

- Support the infrastructure for STI
- Promote the diversification of the national economy by linking research outputs with companies towards industrial diversification
- Capacity building and contributing to the training of the younger generations for the labour market
- Raise scientific awareness among society segments to facilitate the transition towards a knowledge-based society

The goals of NSTIP can be achieved through nine strategic programmes, namely

- Basic research programmes
- Innovation research programme

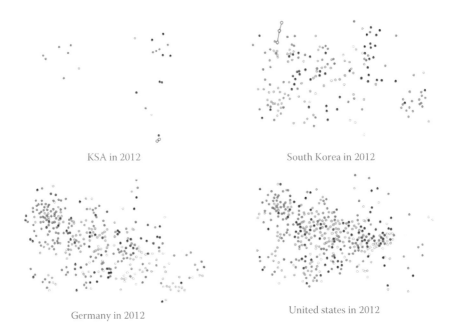

FIGURE 17.4 Illustration of the products of Saudi Arabia in the year 2012 compared to South Korea, Germany and the United States.

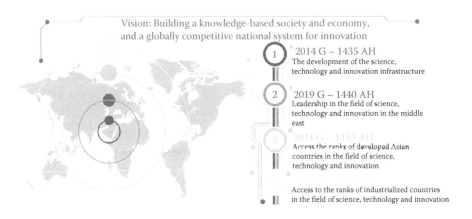

FIGURE 17.5 Pictorial representation of the long-term plans of KACST to achieve the vision for Saudi Arabia.

- National projects programme
- Research and development tools programme
- Innovation tools and technology transfer programme
- Human resources programme
- Science, technology and society programme

- Financial resources development programme
- Management and government's programme

The science and technology research and productivity are achieved through innovations. KACST adopts an efficient innovation cycle to support innovative ideas ranging from basic research to developing high value-added products. The cycle consists of four 'D' phases, namely

- Discover (research)
- Develop (strategies, procedure)
- Deliver (manufacturability)
- Deploy (production and marketing)

SCIENTIFIC AWARENESS OUTREACH PROGRAMME: SCIENCE FOR ALL

KACST aims to promote 'scientifically aware' citizens with favourable behaviours towards science-related fields. The mechanism for the development of scientific awareness is through scientific translation from different languages to Arabic language. It was a national effort to enrich the Arabic content of science and technology on the web.

UNESCO defines a 'scientifically aware person' as one who possesses the characteristics listed in Table 17.1.

Focus Groups

Saudi citizens express a desire for greater exposure to science and technology. The scientific awareness outreach programmes are focused on four groups classified as

TABLE 17.1

Characteristics and Definition of a 'Scientifically Aware Person'

Characteristics	Definition
Knowledge	Knowing the role of science in society and appreciating the cultural conditions accompanying its development
	Having an understanding of the basic conceptual inventions and investigative procedures
Appreciation	Appreciating the role of science in a humanistic way; and feeling comfortable when reading or talking with others about science at a non-technical level
Curiosity	Being curious about the how's and why's of materials and events and being genuinely interested in hearing and reading about things that claim the time and attention of scientists
Openness	Being conversant with the ideas that are being considered in the fields of science without necessarily having to create any ideas pertaining to science
Holistic view	Understanding the interrelationships of science and society, ethics, the nature of science, including basic concepts and relationships of science and humanities

adult males, adult females, university students and high school students. The groups have selected three insights to focus on, namely

1. Positive perceptions which are demonstrated by
   a. High priority
   b. Source of advancement and power
   c. Source of well-being for citizens
   d. In line with religion
   e. Prestigious and noble
2. Challenging perceptions which are defined as
   a. Unclear distinction of science fields
   b. Complicated and rapid pace of change
   c. Lack of government support (progress associated with private sector)
   d. Insufficient communication with science institutions
   e. Stagnant educations system
   f. Lack of broad perspective on scientific careers
   g. Limited visibility on practical applications
3. Expectations that they want to achieve:
   a. Expanded and quality exposure to science
   b. Development of the human element
   c. More tangible experience of science
   d. Integrated national action plan
   e. Desire for varied and interactive channels
   f. Government and private support

COLLABORATING INSTITUTIONS

KACST Science and Technology Outreach programme supports existing institutions with high quality projects and with similar targets. They are the Ministry of Education, Ministry of Higher Education, King Fahad University of Petroleum and Minerals, National Center for Assessment in Higher Education, General Presidency for Youth Welfare, Ministry of Culture and Information and King Abdulaziz University for Science and Technology. The support programme paradigm for the collaborating institutions is shown in Figure 17.6. KACST and the collaborating institutions cater to 10 stakeholders of varied interests and programmes in terms of science and technology. There are: general public, schools and universities, government sector, scientific associations, academe, research institutes, private sector, investors and corporations, decision makers, publishers of scientific publications, press and media, and international stakeholders.

OUTREACH PROJECTS

Acquisition of Printed Materials

The mechanism for the outreach programme should be dynamic and continuing. KACST uses the print media as one of the project tools. Available are the Journals

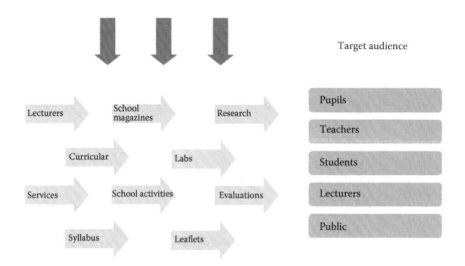

FIGURE 17.6 Schematic representation of the paradigm for support of the collaboration institutions to the target audience.

of Strategic Technology where KACST has Springer as its partner for publication of international journals to foster the development of key applied technologies, providing a forum for the dissemination of research advances and successes. KACST also continues to acquire 77 journals of science and technology and the Arabic version of Nature magazine. Focusing on the interests of teens, 12 journals are available for them.

Multimedia

Discovery is an innovation cycle that is a very effective tool for research and technology quests. KACST has multimedia production of videos, audio and interactive media. It provides scientific multimedia content and they are available to all focus group communities on the Internet (KACST Channel, www.youtube.com/user/kacstchannel; and science talk, www.soundcloud.com/kacst).

To foster success, KACST launches the communications and networks awareness and provides web access to scientific databases, scientific magazines and journals and education publications. Partner organisations for this project are UNESCO, Nature publishing group, Springer, Google, Larousse, Arab Thought Foundation and Learners.

The scientific databases that were established and are continuously updated are shown in Table 17.2.

Future Scientist Project

KACST stimulates young students to be scientists in the future by encouraging then to be scientifically aware. These young students from high school to college levels can have access to http://futurescinetists.kacst.sa to broaden their knowledge, develop skill and gain the attitude towards research.

TABLE 17.2
Database of Scientific Materials

Database Content	Quantity
Theses	205,000 titles
Arabic national documents	65,000
English national documents	91,000
Saudi Automatic Bank for Scientific Terminology (BASM) terms	700,000
Dictionaries	3,000
Scientific journals	60,000
Scientific books (in library)	28,000
Arabic scientific books	53,000
Digitisation of articles	560,000

Messages (20 min)

KACST has developed 20-min interesting and simple messages to capture the interest of children, teens, adults and general public on science and technology. These messages are displayed on outdoor screens, indoor screens, as well as on YouTube and the KACST website.

KACST SCIENTIFIC TRANSLATION EFFORTS

The scientific community of Saudi Arabia needs to be abreast of research findings, ongoing research projects and current updates on research and technology from countries around the world. To meet this need, there is a very active technical arm that undertakes scientific translation of journals, magazines and books. The output of the translation effort is very satisfactory and they are listed in Table 17.3.

TABLE 17.3
Scientific Translation Output

Journal/Magazine Book	Translation Type	Quantity
Nature Magazine	English to Arabic	44 issues
Science and Technology for Teens Science & Vie and Science and Vie Junior	French to Arabic	15 issues
Mathematics of Planet Earth	English to French to Arabic	13 articles
Experiencing Mathematics	English to Arabic	200 lessons and experiments
Strategic Technologies (books)	Arabic to English	45 references 15 technologies
Scientific Culture for All Focus Groups	English/French to Arabic English to Arabic	More than 100 books

SCIENCE, TECHNOLOGY AND INNOVATION
KEY PERFORMANCE INDICATOR (KPI)

The success indicators for the STI programme of KACST are expressed in terms of the key performance indicators (KPI). The KPI values are as follows:

KPI 1: Ranking in scientific publishing quality

Saudi Arabia ranked 35 in the world ranking for the classified journals by Nature magazine. Figure 17.7 shows the result of the Nature survey.

KPI 2: National research output among leading research economies in the Arabian, Persian and Turkish Middle East countries

Saudi Arabia has a steep increase in publication output from the year 2009 to 2012. The output in the year 2012 was 9,000, as shown in Figure 17.8.

KPI 3: Average citation impact (for scientific publications) of Saudi Arabia compared with other countries:

The Normalized Citation Impact (NCI) for Saudi Arabia showed a steep increase from 0.5 in year 2004–2008 to 0.95 in 2012–2013 as reported by Thomson Routers Web of Science. The national annual data of 10 countries shown in Figure 17.9 were benchmarked against a global background for the indicated year.

KPI 4: Patent output of public universities and research institutions

The patent output for the public universities and research institutions totalling 30 gained a peak in the year 2011 with a total number of 260 patents approved as shown in Figure 17.10. This figure was based on the report of Thomson Reuters Web of Science.

Rank	Country	# of scientific papers	2013 partial weighted #	2012 partial weighted #	Change percentage between 2012–2013
1	United States	27,355	18,786,65	18,786,65	−0.8%
2	China	7,637	5,205,60	4,528,97	14.9%
3	Germany	7,669	4,076,97	4,038,30	1.0%
4	Japan	5,102	3,370,85	3,451,26	−2.3%
5	United Kingdom	7,373	3,290,35	3,259,46	0.9%
35	Kingdom of Saudi Arabia	288	76,64	52,52	45.9%
36	Chile	717	75,52	71,91	5.0%
37	Iran	121	58,55	71,45	18.1%
38	Turkey	202	57,07	57,79	−2.9%
51	Egypt	80	12,04	7,32	64.6%

FIGURE 17.7 Ranking of different countries in scientific publishing quality. Saudi Arabia ranks 35th.

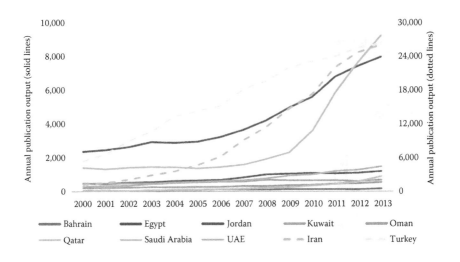

FIGURE 17.8 Research outputs of the Arabian, Persian and Turkish countries in the Middle East from 2000 to 2013.

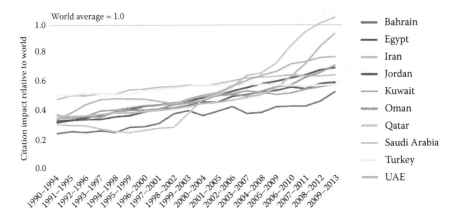

FIGURE 17.9 Annual citation impact of the ten countries from 1990 to 2013.

> ***KPI 5:*** Ranking in the Arabic and Islamic countries that obtained patent approval from the U.S. Patent Office

On the basis of the report of the U.S. Patent Office, Saudi Arabia ranked second among ten Arabic and Islamic countries that were granted approval by this Office as shown in Figure 17.11. It is a great success for Saudi Arabia to rank next to Malaysia which is known to be a technologically advanced country.

KACST LIGHT HIGHLIGHTS

The source of light is important in man's daily living, education, research and technology development, as well as in medicine and all the other facets of life. Saudi

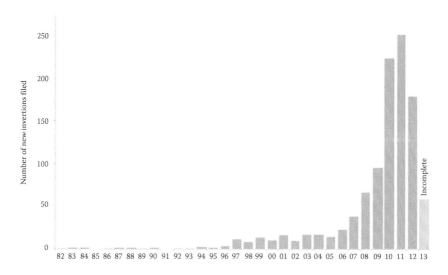

FIGURE 17.10   Number of approved patents per year submitted by 30 public universities and research institutions in Saudi Arabia.

#	Country	Year					
		2010	2011	2012	2013	2014	Total
1	Malaysia	224	181	219	230	271	1125
2	Saudi Arabia	58	61	173	239	294	825
3	Turkey	45	52	55	83	103	338
4	Kuwait	14	24	33	86	101	258
5	Egypt	20	21	28	34	45	148
6	UAE	9	13	21	19	60	122
7	Iran	8	17	28	40	29	122
8	Indonesia	6	11	12	15	15	59
9	Lebanon	5	21	8	9	8	51
10	Pakistan	2	4	13	12	10	41

FIGURE 17.11   Patent ranking of Arabic and Islamic countries that were reported by the U.S. Patent Office.

Arabia through national laboratories and fund agencies has developed programmes to meet this need. It has the Solar Energy programme that is aimed to utilize its optimum solar radiation potentials (based on its location and climate) to invest in solar energy as source of renewable energy. It has developed and continues to improve its solar panel production line and solar panel testing laboratory. Saudi Arabia has a national programme for electronics, communications and photonics that is becoming promising for its development. It has also started to connect cities and homes for communication using fibre optics. With the vast area of Saudi Arabia, which is almost three times the area of France, the great coverage for fibre optic connection is a benchmark.

RECOMMENDATIONS

Saudi Arabia is in the age of infancy in STI. However, its growth in development in these areas is very fast. The national strategic plans of the country have given remarkable outputs, and these outputs surpass those of other countries. International standards and regulations should be observed to maintain the high level of achievement in STI. It is important to maintain and strengthen its international efforts for translation across the languages. International standards and regulations implementation should be maintained and kept in place to safeguard the different projects and programmes. To carry out these programmes, there should be sustainable initiatives to maintain low cost of resources, free access to international databases, less human intervention and high quality standards. Dissemination of knowledge will greatly help transfer societies from shadow to light. Let that quest for knowledge not cease.

REFERENCES

1. Harrub, B, Thompson, B, Miller, D 2003. The origin of language and communication. *TJ*, 17, 93–101.
2. Montgomery, S 2000. The science of translation. In: *Movement of Knowledge through Cultures and Time*. University Chicago Press, USA.
3. School of Edessa. https://en.wikipedia.org/wiki/School_of_Edessa
4. Gundeshapur. https://en.wikipedia.org/wiki/Gundeshapur
5. House of Wisdom. https://en.wikipedia.org/wiki/House_of_Wisdom
6. Mrayati, M, Allan, Y M, At-Tayyar M H 2002. Series on Arabic origins of cryptology. In: *Al-Kindi's Treatise on Cryptoanalysis*. KFCRIS & KACST, Riyadh, Saudi Arabia.

18 Optics and Photonics Research in Lebanon
Key Figures, Ways Forward

Marie Abboud Mehanna

CONTENTS

Regional and Historical Context 243
National Context 244
Challenges 246
Ways Forward 246
References 247

REGIONAL AND HISTORICAL CONTEXT

Lebanon's location at the crossroads of the Mediterranean Basin and the Arabian hinterland facilitated its rich history and shaped a cultural identity of religious and ethnic diversity. The Lebanese are descended from many different peoples who have occupied, invaded or settled in this corner of the world, making Lebanon a mosaic of closely interrelated cultures. Lebanon was and still is an intermediary among various civilisations and cultures. Perhaps the clearest indication of its link between the East and the West is its role in the Arab renaissance of the nineteenth century which revitalised and modernised the intellectual movement, founded schools, universities, presses, newspapers and magazines, and initiated research and translation [1]. The modern education base of Lebanon was structured by missionaries who came to the country centuries back for the purpose of reinforcing the Catholic Christian communities through spreading education. The American University of Beirut (AUB) and the Saint-Joseph University of Beirut (USJ) were the first Anglophone and the first Francophone universities to open in Lebanon respectively. The establishment of formal higher education in Lebanon began with the founding of the Syrian Evangelical College by the American Evangelical Mission in 1866, which in 1920 came to be known as the AUB. In 1883, the Society of Jesus founded Saint Joseph University; this university was a branch of the University of Lyon in France, and it gained its independence in 1975. The Jesuits played a notable role in these activities. In 1839, they came to Beirut and established a modest school. This was followed in 1855 by a school in Ghazir (Kisruwan) which would be moved to Beirut in 1875. Authorities graced the new school with the title of 'university' which allowed it to grant academic degrees.

The establishment of the observatory of Ksara by the fathers of the Society of Jesus in 1911 was an Oriental response to the industrial and scientific activity in Europe [2]. The beginnings of the Ksara Observatory were very modest. As it was in January 1911, the observatory of Ksara included meteorology, geomagnetism, seismology and astronomy. Its activity was mainly reduced to provide very accurate time and location.

Nowadays, while the Lebanese educational system offers a high quality education, the local employment market lacks sufficient opportunities, encouraging many of the young people to leave the country to work abroad and become the business backbone of regional and international markets. An overwhelming majority of emigrants (77%) are young people under 35 years, many of whom are well-educated and highly skilled professionals who exacerbate the loss to Lebanon in terms of economic and social advancement. Indeed, 44% of those who emigrated held university degrees [3]. This brain drain phenomenon is not an issue of the current century as it started since the early 1900s. Hassan Kamel al-Sabbah, the Oriental Edison [4,5] is an example of a young scientist who moved to the United States of America. Al-Sabbah worked at the Engineering Laboratory of General Electric Company in New York after signing a contract which stated that all his inventions were owned by the company. Having around 70 U.S. and foreign patents covering his work, he received a reward of 1 USD for each of his patented inventions. In 1935, a few months before the fatal car accident, Al-Sabbah declared that he would return to the Middle East and transform the Arabian Desert into a paradise. Since the main ingredients for solar power are the ubiquitous sand which is used to make solar cells, and the strong sun for powering them, Al-Sabbah intended to use the desert to generate and power the solar cells, thus producing enormous amounts of energy.

NATIONAL CONTEXT

Academic: Like in the majority of Arab countries, the bulk of scientific production in Lebanon is based in and issued from higher education institutions. Lebanon has 45 universities or higher education institutions, several of which are internationally recognised. The universities, both public and private, largely operate in French or English as these are the most widely used foreign languages in Lebanon. According to the Webometrics Ranking of World Universities (http://www.webometrics.info), the top-ranking universities in the country are the AUB, the USJ, the Lebanese University (founded in 1951), the Lebanese American University and the Holy Spirit University of Kaslik. Six Lebanese universities are listed in the top 50 Arab universities according to the most recent study performed by the British society Quacquarelli Symonds [6].

As for academic programmes in the physical sciences, only ten universities offer a programme leading to the degree of BS in physics while five offer various programmes leading to the degree of MS in physics (Physics of sensors and instrumentation; Astrophysics; Laser and Material; Nanosciences and Functional Materials; Energetics; Medical Physics; Condensed Matter Physics; Physics (condensed matter, material science, high energy physics); Nuclear physics and Radiation protection; Advanced physics). Although the master degree 'Laser and Material' provided by the Lebanese University is the only degree that can be indexed in the field of optics and photonics, other MSc programs offer modules in photonics courses, broadening the spectrum of applications from the interaction between matter and light to opto-electronics. For the PhD degree, five universities offer Doctorate degrees in selected domains of Science & Technology and/or Engineering.

Research activities: Lebanon has been exhibiting different research specialisation trends and various growth rates in producing scientific outcomes in recent years. In optics and photonics, research includes all their related aspects of light generation, detection and manipulation, through emission, transmission, modulation, amplification and detection. A presentation of all research activities in photonics would be an impossible task since some activities are conducted as a researcher's personal initiative. Therefore, I limit my presentation in this article to a few leading and innovative examples of current research projects carried out by Lebanese researchers, by apportioning them according to the researcher's academic affiliation.

At USJ, research projects range from wave propagation to light-matter interaction using ionising radiations or optical imaging, including experimental, theoretical and numerical aspects. A first axis concerns optical metrology in scattering media and, more particularly, polarimetric and coherent aspects in biophotonics. Emphasise is attributed to applications in life sciences, covering applications in metrology [7], environmental [8] and medical [9,10] fields as well as food industry [11,12]. A second theme deals with ionising radiations for radiobiology and radiotherapy issues [13–15]. Improvements to cancer treatment planning, implementing and assessing techniques to precisely target beams of radiation are continuously performed. A third example is related to propagation characteristics of electromagnetic waves in various media (tunnels, waveguides …). Numerical simulation methods, such as domain decomposition or method of integral equations, are used. Cases dealing with the diffraction of electromagnetic waves by different types of barriers, the transmission of electromagnetic waveguides are considered [16,17].

At the Lebanese University, the only public university in Lebanon, the platform for research in nanosciences and nanotechnology has been recently equipped to serve research needs in many disciplines: materials sciences, chemical physics, electronics and microelectronics, dealing with a broad spectrum of applications [18–22].

At the American University of Beirut, and among many research domains, interest and expertise lie in thin film physics, growth and characterization; particularly laser and plasma assisted growth of functional materials, using techniques such as pulsed laser deposition, pulsed laser melting and microwave plasma assisted deposition to investigate the synthesis of a wide variety of technologically important and scientifically challenging materials, ranging from oxides, nitrides, carbon based materials to chalcogen-supersaturated silicon, these latter showing sub-band-gap infrared absorption with potential applications in solar cells and infra-red detectors [23,24]. Development, synthesis, spectroscopic and electronic characterization of materials for optoelectronic device applications are also conducted at the Lebanese American University [25,26] where researchers employ new materials in the preparation and development of Organic Light-Emitting Diodes.

Research funds: According to statistical data in the Arab Knowledge Report of 2009 [27], the total investment of the entire Arab world in research and development is a meagre 0.2%–0.3% of its Gross Domestic Product compared to 2.0%–4.9% in the United Kingdom, Germany, Sweden, Israel, Japan and the United States of America, individually. Also, as opposed to other parts of the world where the private sector plays a significant role, most Arab countries depend on government funding for scientific research and very weak links exist between research institutes and production sectors.

In Lebanon, the main funds for research are provided by: (i) government funding for all scientific research, with the Lebanese National Council for Scientific Research (CNRS-L) playing a major role at the national scene, offering diverse funding schemes and opportunities [28]; (ii) international agreements or institutions; examples are given by the CEDRE programme; EU programmes that include Mediterranean partner countries (FP7; ERANETMED; H2020) and UN programmes and (iii) research boards and councils of private universities, for instance the Conseil de la Recherche at USJ and the Research Board at AUB.

CHALLENGES

Lebanon faces a host of hurdles when it comes to higher education and scientific research. Challenges can be declined at the scales of both the entire nation and the higher education system.

National level: Brain drain is a major issue that depresses investment in S&T capacity as the return to this investment is lost as soon as the trained population moves abroad. Data indicate that about one-third of qualified scientists and engineers who were born in developing countries move to developed nations to work [29]. This average is higher in Lebanon (43.9%) [3]. Reasons for the brain drain vary, but generally include low salaries, lack of basic instrumentation and technical support, lack of access to high level research networks, dismal lack of research opportunities, highly uncertain socio-economic conditions for the future, and weak integration of basic science and technology or R&D with public and private enterprises [30].

Higher education institution: Higher education institutions often adopt the method of transferring and memorising knowledge rather than getting it through research. This can be illustrated by scientific research of graduate students that is rather traditional and rarely tackles socio-economic development, except for a few timid initiatives. Furthermore, scientific research and national sustainable development plans are totally disconnected. Finally, research results are unemployed in economic projects because of the lack of strong relation between research institutes and production sectors.

WAYS FORWARD

It seems that the only clear solutions would be to increase the budget for scientific research, select meaningful priority areas for research, lay down workable strategic goals and action plans, establish adequate databases and networking capabilities, and robustly encourage private sector input and participation.

One of the main key issues related to brain drain losses is the mismatch between the highly skilled graduates that universities supply and the labour market demands in terms of skills and professionals. These supply–demand failings in the labour market are increasingly preventing young people from finding a job that fits their qualifications. In order to address these issues, universities can implement policies to reform the curricula to better prepare graduates for their professional life, improve career guidance structures and activities, and invest in relevant research. Any worthwhile relevant research must necessarily be based on universal pillars, vision and strategy, logistics, human resources that include well qualified researchers, and meaningful research priorities directed towards problem-solving rather than just publishing.

In the field of optics and photonics and since light can be used to generate energy or to transfer information, sustainable development with 'Green' Job Opportunities and/or 'Light' Job Opportunities can be foreseen. Pertinent initiatives to the Lebanese scene can be dispensed by examples related to the solar energy development and the communication field. Indeed, in Lebanon, with an average of 300 days of sunshine per year, solar energy for heat production and photovoltaic for electricity production is, to date, less than 1% of the total energy production of the country. The benefits of solar energy would include filling part of the deficit in the country, substantial savings for the Lebanese households, reducing the national fuel bill, and reducing emissions of greenhouse gases. In this regard, the Shaams project [31] implemented under the ENPI CBC Mediterranean Sea Basin Programme aimed at facilitating the take-up of solar technologies by raising public awareness on energy efficiency through the transferability and implementation of good practices in legal, regulatory, economic, organisational issues and new financing mechanisms. The Lebanese partners, Berytech and BIAT incubators, worked with their partners in this Euro-Mediterranean consortium on three fronts: Policy Accelerator, Enterprise and Research Accelerator, and Social Accelerator, in order to lift the political and technical barriers to the development of solar energy.

As per the communication network, implementing a fibre optics network in the country is another endless geopolitical bickering. The fibre optic infrastructure already exists and most, if not all, internet stations in Lebanon are interconnected by fibre optics. The real challenge resides in connecting these stations to organisations, businesses and households. The benefits of the network would include substantial savings for the Lebanese households, fast and abundant internet connections since the current average internet speed is 3.11 Mbps, putting Lebanon back on track, maybe turning it into a tech hub for the MENA region and allowing access to grid infrastructures and facilities for research activities.

REFERENCES

1. Kassir S 2004. *Considérations sur le malheur arabe*. Actes-Sud, Paris.
2. Berloty P 1912. L'Observatoire de Ksara. *Ciel et Terre* 33, 103.
3. Yaacoub N, Badre I. 2011. *The Labor Market in Lebanon*. Statistical report, Central Administration of Statistics, Beirut.
4. Wahbī A 2004. *Al-Faylasūf al-'abqarī Ḥasan Kāmil al-Ṣabbāḥ*. Dār al-Janūb lil-Ṭibā'ah, Beirut.
5. http://www.crdp.org/en/details-edumagazine/117/419
6. http://www.topuniversities.com/university-rankings/arab-region-university-rankings/2015

7. Abou Nader Ch, Pellen F, Roquefort Ph, Aubry T, Le Jeune B, Le Brun G and Abboud M 2016. Evaluation of low viscosity variations in fluids using temporal and spatial analysis of the speckle pattern. *Optics Letters* 41(11): 2521–2524.
8. Nassif R, Abou Nader Ch, Rahbani J, Pellen F, Salameh D, Lteif R, Le Brun G, Le Jeune B, Kallassy M and Abboud M 2015. Characterization of Bacillus thuringiensis parasporal crystals using laser speckle technique: Effect of crystals dimension and concentration. *Applied Optics* 54(12): 3725–3731.
9. Abou Nader Ch, Pellen F, Loutfi H, Mansour R, Le Jeune B, Le Brun G and Abboud M 2016. Early diagnosis of teeth erosion using polarized laser speckle imaging. *Journal of Biomedical Optics* 21(7): 07110301–07110306.
10. Joud F, Warnasooriya N, Bun P, Tessier G, Atlan M, Desbiolles P, Coppey-Moisan M, Abboud M and Gross M 2010. Imaging gold nanoparticles in living cell environments using heterodyne digital holographic microscopy. *Optics Express* 18: 3264–3273.
11. Nassif R, Pellen F, Magné C, Le Jeune B, Le Brun G and Abboud M 2012. Scattering through fruits during ripening: Laser speckle technique correlated to biochemical and fluorescence measurements. *Optics Express* 20: 23887–23897.
12. Nassif R, Abou Nader Ch, Afif Ch, Pellen F, Le Brun G, Le Jeune B and Abboud M 2014. Detection of Golden apples climacteric peak by laser biospeckle measurements. *Applied Optics* 53(35): 8276–8282.
13. Francis Z, Seif E, Incerti S, Champion C, Karamitros M, Bernal MA, Ivanchenko VN, Mantero A, Tran HN and El Bitar Z 2014. Carbon ion fragmentation effects on the nanometric level behind the Bragg peak depth. *Physics in Medicine and Biology* 59(24): 7691–7702.
14. Francis Z, Incerti S, Ivanchenko V, Champion C, Karamitros M, Bernal MA and El Bitar Z 2012. Monte Carlo simulation of energy-deposit clustering for ions of the same LET in liquid water. *Physics in Medicine and Biology* 57(1): 209–224.
15. Farah N, Francis Z and Abboud M 2014. Analysis of the EBT3 Gafchromic film irradiated with 6 MV photons and 6 MeV electrons using reflective mode scanners. *Physica Medica: European Journal of Medical Physics* 30(6): 708–712.
16. Sayah T and Azar B 2004. Effect of junction on electromagnetic waveguides transmission. *Mathematica Balkanica* 18.
17. Sayah T and Azar B 2003. Coupling finite element method-modal expansion method for the junction of electromagnetic waveguides. *International Journal of Applied Mathematics* 12(3).
18. Maurer T, Nicolas R, Lévêque G, Subramanian P, Proust J, Béal J, Schuermans S et al. 2014. Enhancing LSPR sensitivity of Au gratings through graphene coupling to Au film. *Plasmonics* 9(3): 507–512.
19. Herro Z, Zhuang D, Schlesser R and Sitar Z 2010. Growth of AlN single crystalline boules. *Journal of Crystal Growth* 312(18): 2519–2521.
20. Zaouk D, Al Asmar R, Podlecki J, Zaatar Y, Khoury A and Foucaran A 2007. X-ray diffraction studies of electrostatic sprayed SnO_2: F films. *Microelectronics Journal* 38(8): 884–887.
21. Zaatar Y, Zaouk D, Bechara J, Khoury A, Llinaress C and Charles J-P 2000. Fabrication and characterization of an evanescent wave fiber optic sensor for air pollution control. *Materials Science and Engineering B* 74(1): 296–298.
22. Zaatar Y, Bechara J, Khoury A, Zaouk D and Charles J-P 2000. Diode laser sensor for process control and environmental monitoring. *Applied Energy* 65(1): 107–113.
23. Isber S, Skaff N, Roumie M and Tabbal M 2014. Growth of Co_2MnAl thin films by pulsed laser deposition. *European Journal of Physics: Web of Conferences* 75: 03008.
24. Kassem W, Tabbal M and Roumie M 2011. Pulsed laser deposition of tungsten thin films on graphite. *Advanced Materials Research* 324: 77–80.

25. Wex B, Jradi FM, Patra D and Kaafarani BR 2010. End-capping of conjugated thio-phene-benzene aromatic systems. *Tetrahedron* 66: 8778–8784.
26. El-Assaad T, Auer M, Castañeda R, Hallal KM, Jradi FM, Mosca L, Khnayzer RS et al. 2016. Tetraaryl pyrenes: Photophysical properties, computational studies, crystal structures, and application in OLEDs. *J. Materials Chemistry. C* 4: 3041–3058.
27. Arab Knowledge Report. 2009. *Towards Productive Intercommunication for Knowledge.* UNDP.
28. http://www.cnrs.edu.lb/
29. Meyer JB 2003. *Policy implications of the Brain Drain's changing face*, Science and Development Network, Brain Drain, policy briefs.http://www.scidev.net/dossiers/index.cfm?fuseaction=printarticle&dossier=10&policy=24 (2.12.2006).
30. World Bank. 2013. *Transforming Arab Economies: Travelling the Knowledge and Innovation Road*, World Bank, Washington.
31. http://www.shaams.org/

Index

A

Accidental colors, *see* After-images
After-images, 76; *see also* Color science
Age of the Enlightenment, 21
Al-Basri, *see* Ibn al-Haytham
Al-Bitrūjī, 57
Alhacen, *see* Ibn al-Haytham
Al-Ḥākim, 10, 11
Al-Hassan Ibn al-Hassan Ibn al-Haytham,
        see Ibn al-Haytham
Al Hazen, *see* Ibn al-Haytham
Almagest theory, 60
Al-Mu'izz, 10
AlN on Si, growth of, 160
Al-Shukūk, 58, 61–62
American University of Beirut (AUB), 243
Amorphous material, 160
Analytical Mechanics, 37, 38
Anthropomorphism, 17
APCTP, *see* Asia Pacific Center for Theoretical
        Physics
Arabic-Islamic tradition, 54
Arabic period, renewal of scientific studies in, 57
Aristotle
        the substances and the accidents, 18
        universality, 14
Asia Pacific Center for Theoretical Physics
        (APCTP), 213
Astronomy, 53, 54; *see also* Duhem; Ibn
        al-Haytham
        al-Bitrūjī, 57
        conceptions of, 56
        Copernican revolution, 53–55
        types of approach to, 56
Atoms' motion inside molecules, 167
AUB, *see* American University of Beirut
Avicenna, 13

B

Band widening, 140
Bank for Scientific Terminology (BASM), 238
Barkla, Charles Glover, 166
Basic necessities, 207
BASM, *see* Bank for Scientific Terminology
Baṣra, 11
Bayhaqī, 11
Beamlines, 206, 221, 224
Biconvex hyperbolic lens, 196

Binocular vision, 82, 145; *see also* Physiological
        optics
        corresponding points, 86
        crossed and uncrossed visual directions, 85
        cyclopean axis, 84
        experiments, 83
        Hering's law of visual directions, 84
        stereoscopic vision, 84
Biophotonics, 137
Bīrūnī, 13
Blue LED, 15
Bohr, Niels, 34
        complementarity, 35
Book of Optics, The, 6–8, 26; *see also* Ibn
        al-Haytham's research programme
        conversio, 47
        Latin translation, 44–45
        obliquatio, 47
        reflection and refraction, 45, 47, 48, 49
        reflexio, 45–47
        refraction, 45
        refraction of object through sphere
                of glass, 44
        terms and ambiguity, 49–50
        translation and interpretation, 43, 50–51
Bose–Einstein statistics, 34–35
Bosons, 34
Bruno, Giordano, 13
Bundle theory, 18

C

Cairo, 10
        House of Wisdom, 11
Camera, 161
Camera obscura, 72–73, 161; *see also*
        Physiological optics
        hole size and image quality, 162
Catoptricians, 4
CERN, *see* Conseil Européen pour la Recherche
        Nucléaire
CESSAMag, 227
ChemCam, *see* Chemistry Camera
Chemistry Camera (ChemCam), 143
        acquired remote laser induced breakdown
                spectra, 146
        basis of, 144
        LIBS, 144, 145
        plasma plume, 145
        RMI camera, 145

CNRS-L, *see* Lebanese National Council for
    Scientific Research
Cognitive process, 18
Color science, 73; *see also* Physiological optics
  after-images, 76
  color constancy, 78
  color contrast, 77
  color wheel, 74–75
  constancies, 76–77
  illumination, 74
  inference, 73
  loss of color in peripheral visual field, 74
  visual system and color, 74
Color wheel, 74–75
Colours of visible light, 165
Communication, 229
Compact fluorescent lamp, 184
Complementarity, 35
Concavity, 80
Conical instruments, 136
Connected LED street lighting, 187–188
Connected lighting system, 188–190
Conseil Européen pour la Recherche Nucléaire
    (CERN), 210
Contracting Party countries, 210
Conventional lighting products, 184
Convergence law, 89
Conversio, 47; *see also Book of Optics, The*
Copernican revolution, 30, 53–55

D

D'ALEMBERT, Jean, 37
Darkness, 15
Debye–Scherrer scattering, 166
Deflection of light, 136
Demythologisation, 21
Dense Wavelength Division Multiplexing
    (DWDM), 139
Depth perception, 146
Descartes, René, 36, 37
DESY, *see* Deutsches Elektronen-Synchrotron
Deutsches Elektronen-Synchrotron
    (DESY), 224
Dioptrics, 8
Diplopia, 82, 89; *see also* Binocular vision
Directionally sensitive, 79–80
Disc of least confusion (DLC), 157
Discovery, 237
DLC, *see* Disc of least confusion
Double refraction effect, 147
Dualistic model, 15
Duhem, 54–55; *see also* Ibn al-Haytham
  Arabic astronomical tradition
    interpretation, 55
  about atomist theory, 56
  conceptions of astronomy, 56

  conflict as being philosopher and historian, 58
  criticism of Ptolemaic system, 57
  description of state of astronomy, 56–57
  as natural philosopher, 55
  Newtonian physics, 55
DWDM, *see* Dense Wavelength Division
    Multiplexing

E

Einstein, 30
  Bose–Einstein statistics, 34–35
  bosons, 34
  conception of physical principles, 32
  contributions in quantum domain, 34
  discrete nature of energy of radiation, 34
  domain of material reality, 33
  First Quantum Theory, 34
  fluctuation calculation, 35
  General Theory of Relativity, 33
  principles, 31–32
  principles theories, 35
  space-time, 32
  Special Theory of Relativity, 32–33
  theory, 31
Electric light, 182–184
Electromagnetic radiation, 166
Electron
  diffraction and scattering, 166
  energy loss spectroscopy, 167
  microscope, 132
  –solid interaction, 154, 155
Electronic-photonics convergence, 138
Electron microscopy, 158, 159
  as fundamental research tool, 154
Emission theory, 16
Energy of radiation, 34
Equal innervation law, 87–88
Erazmus Ciołek Witelo, 104
ESA, *see* European Space Agency
ESF, *see* European Science Foundation
ESRF, *see* European Synchrotron Radiation
    Facility
Euclidean continuum, 127
European science, 209–211; *see also*
    International Science Centres in Arab
    Countries
  CERN, 210
  Contracting Party countries, 210
  joint initiatives in Europe, 210
  Learned Societies, 209
European Science Foundation (ESF), 215
European Space Agency (ESA), 210
European Synchrotron Radiation Facility
    (ESRF), 170, 198, 210, 226
EXAFS, *see* Extended x-ray absorption fine
    structure

Existence of object, 18
Extended x-ray absorption fine structure
        (EXAFS), 166
Extramission theories, 124, 194
Extromissionism, 70
Eye, 68; *see also* Physiological optics
    eyeball, 81
    model, 69

F

Fechner, Gustav Theodor, 67
Femtochemistry, 167
Femtosecond pump-probe spectroscopy, 167
Femtoseconds (fs), 167
Femtosecond x-rays, 171, 172
Fermi-Dirac statistics, 35
Fiber to the x, *see* FTTx
First Quantum Theory, 34
Flectere, 46
Frangere, 46
fs, *see* Femtoseconds
FTTx (Fiber to the x), 139, 140
    development of, 140–141
Fundamentalism, 12

G

Galen, 194
Gallium nitride (GaN), 159
General Theory of Relativity, 33; *see also*
        Einstein
GNP, *see* Gross national product
God
    mystical union with, 19
    rationalist theology, 17–18
    structure, 17
    vision, 17
Gross national product (GNP), 208
Ground theory hypothesis, 98
Guided optics, 135
Gullstrand, Allvar, 69

H

Halogen lamp, 184
Helmholtz, Herman von, 64
Hering's law of visual directions, 84
Herschel, Sir Frederick William, 165
    heat in colour, 165
HHG sources, *see* High harmonic generation
        sources
HID lamps, *see* High intensity discharge lamps
High harmonic generation sources (HHG
        sources), 169
High intensity discharge lamps
        (HID lamps), 184

High-speed telecommunications network, 139
Hishām b. al-Ḥakam, 17
Horopter, 86–87; *see also* Binocular vision
House of Wisdom, 11
    foundation of, 230
Human visual system, 69; *see also* Physiological
        optics

I

IAEA, *see* International Atomic Energy
        Agency
Ibn al-Hassan, Mohammad, 152
Ibn al-Haytham, 3, 10, 136, 143, 152, 182, 197,
        221; *see also Book of Optics, The*;
        Chemistry Camera; Duhem
    Almagest theory, 60
    al-Shukūk, 58, 61–62
    and Aristotle's universality, 14
    astronomy, 54
    binocular vision, 145
    camera obscura, 161
    celebration of his life and work, 53
    colours visible light, 165
    conception of light, 28
    confusion about name, 153
    contribution in electron microscopy, 159
    demystifier of Copernican Revolution,
        30, 53–55
    demythologisation, 21
    depth perception, 146
    discoveries of, 132
    double refraction effect, 147
    early life, 11
    experimental supply, 12
    explanation of vision, 145
    imprisonment issue, 153–154
    lifestyle, 11
    about light, 14, 21–22
    logical-epistemic basis of astronomical
        pluralism, 58
    loss of light intensity, 146
    Muslim scholars' contribution, 148, 152
    on optical reflection, 146–147
    optics, 54
    physical optics, 25–30
    predecessors, 27
    Ptolemaic assumptions, 59
    Ptolemaic hypotheses, 60–61
    Ptolemy's moon model, 59
    about rationalist theology, 19
    recognition in United Nations, 199
    reflecting surface and image quality, 147
    scientific method of enquiry by, 160–161
    spherical aberration, 157, 162
    study of light optics, 151
    visual ray, 27, 28

Ibn al-Haytham's problem, 109
  Barrow's solution, 115
  bifurcation set, 118
  in *Book on Optics*, 109
  complex solution, 116–117
  geometry of, 114
  hyperbola of Ibn al-Haytham and
        Huygens, 116
  lemma 7.1, 110
  lemma 7.2, 110–112
  ray-tracing solution, 119–121
  solutions of, 112
  trigonometric solution, 113–115
  Tychsen's asteroid, 119
  Tychsen solution, 117–119
Ibn al-Haytham's research programme, 3
  Book of Optics, 6–8
  catoptrics, 7
  characterisation, 4
  conditions for vision, 6–7
  dioptrics, 8
  Ibn al-Haytham's optics, 4
  image of object by refraction, 8
  mathematical study of curve, 5
  Snellius' law, 5
  spherical aberration, 8
  spherical lens, 8
Ibn al-Muqaffa', 15
Ibn al-Rāwandī, 20
Ibn Sahl, Abu Alla Al Sad, 27, 136, 155, 193; *see
        also* Ibn al-Haytham
  biconvex hyperbolic lens, 196
  contribution in electron microscopy, 159
  diagram for biconvex hyperbolic lens, 196
  diagrams for refraction of light, 195
  law of refraction, 155, 156
ICTP, *see* International Centre for Theoretical
        Physics
ICTP-SAIFR, *see* ICTP South American
        Institute for Fundamental Research
ICTP South American Institute for Fundamental
        Research (ICTP-SAIFR), 213
Incandescent lamp, 184
Indoor positioning systems, 190
Infrared (IR), 227
International Atomic Energy Agency (IAEA),
        212, 226
International Centre for Theoretical Physics
        (ICTP), 212–213
International School of Advanced Study
        (SISSA), 212
International Science Centres in Arab Countries,
        207, 215
  Arab scholars and developments, 209
  CERN, 210
  Contracting Party countries, 210
  International Year of Light, 213–215

IWG, 216
  joint initiatives in Europe, 210
  Learned Societies, 209
  LHiSA, 217
  reconstruction of European science, 209–211
  regional science centres, 211–213
  science and nature, 216
  SESAME Project, 211
  statistical data for regions 2014, 216
  Zewail, Ahmed, 208
International Working Group (IWG), 216
International Year of Light (IYL), 135, 213;
        see also International Science
        Centres in Arab Countries
  national nodes of, 214
Intromission
  of forms, 27
  theories, 79, 124, 193
IR, *see* Infrared
Iraqi theologians, 21
Islam, 20
IWG, *see* International Working Group
IYL, *see* International Year of Light

J

Joint initiatives in Europe, 210
Jungk, Robert, 9

K

KACST, 232; *see also* Scientific translation
  innovation cycle, 235
  light highlights, 240
  long-term plans of, 234
  multimedia production, 237
  Science and Technology Outreach
        programme, 236
  scientific translation efforts, 238
  scientific translation output, 238
  Solar Energy programme, 241
Key Performance Indicator (KPI), 239
Kitāb al-Manāzir, see Book of Optics, The
Koyré, Alexander, 62
KPI, *see* Key Performance Indicator
KSA's National Strategy, 231; *see also* Scientific
        translation
  KACST, 232, 234
  NSTIP, 233
  products of Saudi Arabia, 233, 234
  STNP, 232
Kuhn, Thomas, 103

L

Laser, 137
  -driven tabletop plasma sources, 169

Laser-induced breakdown spectroscopy
        (LIBS), 144
Late Antiquity, 18
Law
    of convergence for visual perception, 89
    of equal innervation, 87–88
    of *reflection* of light, 194
    of refraction, 136, 197
LCLS, *see* Linac Coherent Light Source
Learned Societies, 209
Lebanese National Council for Scientific
        Research (CNRS-L), 246
Lebanon, 243
    challenges, 246
    national context, 244–246
    observatory of Ksara, 244
    regional and historical context, 243–244
    research in, 243
    ways forward, 246–247
LED, *see* Light-emitting diode
Lenses, 136
    made from quartz, 193–194
LHiSA, 217
LIBS, *see* Laser-induced breakdown
        spectroscopy
Life, search of, 144
Light, 9, 14, 25, 53, 132; *see also* Optics
    as colour, 18
    conception of, 28
    dualistic model, 15
    focusing of, 136
    and God, 15
    Ibn al-Haytham about, 21–22
    improving energy efficiency, 190–191
    in Genesis, 9
    Iraqi Shiis, 21
    large bandwidth capabilities of light
            signal, 138
    loss of light intensity, 146
    metaphor of, 23
    optics, 151
    as physical and rationality, 23–25
    physical nature of, 28
    principles in technology, 136
    Quran, 14
    and rays, 16
    role in optical telecommunications, 135
    sources, 158
Light-based technologies, application fields, 131
Light-emitting diode (LED), 159
    blue LED, 15
    savings potential from, 186
    search for blue, 159
Lighting, 181, 191
    challenges, 184–186
    connected LED street lighting, 187–188
    connected lighting system, 188–190

conventional lighting products, 184
and display, 138
early impact of, 181–182
electric light, 182–184
eliminating light poverty, 190–191
first central power plant, 183
indoor positioning, 190
LED, 186
milestones in, 182
off-grid solar lighting, 191
Philips Hue ecosystem, 189
Philips Incandescent Lamp Factories, 183
savings potential from LED, 186
zone of uncertainty, 185
Linac Coherent Light Source (LCLS), 173
Listing's law, 80–81
Lorentz' transformations, 31

M

Manipur Renewable Energy Development
        Agency (MANIREDA), 191
MANIREDA, *see* Manipur Renewable Energy
        Development Agency
Manṣūriyya, 10
Mars Science Laboratory (MSL), 143
    mission, 131
Mathematical Analysis, 37
Maxwell's Electromagnetic Theory, 31
Mediaeval Arab achievements in optics, 194,
        198–199
    biconvex hyperbolic lens, 196
    Galen, 194
    Ibn Sahl, 193
    law of refraction, 197
    laws of *reflection* of light, 194
    lenses made from quartz, 193–194
    Medieval scholars, 197
    refraction of light, 195
    SESAME, 198
    Snell's law of refraction, 193
    synchrotron light, 198
Medieval scholars, 197
MESC, *see* Middle East-based Scientific
        Cooperation
Metamaterials, 138–139
Micrometre, 132
Microscopy, 154
Middle East-based Scientific Cooperation
        (MESC), 224
Minkowski, Alexander, 32
Monotheism, 15
Moon illusion, 88; *see also* Physiological optics
    law of convergence for visual perception, 89
    Ptolemy, 89–90
    sight and size, 90–91
Motion of atoms inside molecules, 167

Motion perception, 91; *see also* Physiological
        optics
    induced movement, 93
    reafference principle, 92
    temporal threshold, 91
MSL, *see* Mars Science Laboratory
Muḥammad, 16
Muslim scholars' contribution, 148, 152
Mysticism, 19

N

Nano-photonics, 135
Nanoseconds (ns), 167
Nāṣir-i Khosraw, 12
National Science Technology and Innovation
        Plan (NSTIP), 233
Natural light, 158
Nature
    science and, 216
    timescales of fundamental processes in, 168
Nearest Neighbour technique
        (NN technique), 93
Negative refraction, 139
Newtonian physics, 55
Nimrud/Layard lenses, 194
NN technique, *see* Nearest Neighbour technique
Notion of medium, 27
ns, *see* Nanoseconds
NSTIP, *see* National Science Technology and
        Innovation Plan
Nūrī, 19

O

Obliquatio, 47; *see also* Book of Optics, The
Observatory of Ksara, 244
OED, *see* Oxford English Dictionary
Off-grid solar lighting, 191
OIC, *see* Organization of Islamic Conference
Optical computer, 138
Optical materials, 138–139
Optical reflection, 146–147
Optical science, 3
    branches of, 3
    Ibn al-Haytham's optics, 4
Optical telecommunications, 135
Optics, 28; *see also* Light; Physiological optics
    field of, 63
    physiological, 63
    programme of study in physiological, 64
    science of, 29
Optoelectronic components, 135
Organization of Islamic Conference (OIC), 208
Orthodox, 20
Oxford English Dictionary (OED), 64

P

Pattern classification, 93; *see also* Physiological
        optics
    NN rule, 93
    object recognition, 94
    perception by inference, 94
    reject option decision rule, 96
    role of memory in, 93–94
    template matching, 95
    vision, 94
PC, *see* Photonic crystals
Pendry, Sir John, 139
Perception, 80; *see also* Physiological optics
    active processes, 98
    signs, 100
    stages of vision/perception, 99
    visual, 99
Philips Hue ecosystem, 189
Philips Incandescent Lamp Factories, 183
Photonic crystals (PC), 139
Photonics, 139, 141
    research, 140
Photonic technology, 135, 136, 137, 141
    biophotonics, 137
    diffusing technology, 137
    electronic-photonics convergence, 138
    high-speed telecommunications network
        issues and developments, 136
    large bandwidth capabilities of light
        signal, 138
    laser, 137
    lighting and display, 138
    metamaterials, 138–139
    in military, 138
    nano-photonics, 135
    optical computer, 138
    optical materials, 138–139
    photon in data transmission, 136
    research and jobs, 140
    in space application, 137, 138
    widening of bands, 140
Photon in data transmission, 136
Physiological optics, 63, 103–104; *see also* Color
        science
    anatomy of eye, 67–69
    binocular vision, 82–86
    camera obscura, 72–73
    colour, 73
    concavity, 80
    dark and light adaptation, 78–79
    diplopia, 82
    directional sensitivity, 79–80
    extromissionism, 70–71
    eyeball, 81
    eye movements, 80–82
    glacial sphere, 72

horopter, 86–87
human visual system, 69
intromission theory, 79
law of equal innervation, 87–88
Listing's law, 80–81
moon illusion, 88–91
motion perception, 91–93
object recognition, 94
oeuvre of, 65–67
optics and vision research, 66
pattern classification, 93–96
priming, 100
punctiform analysis, 79
schematic eye models, 69–73
seat of vision, 71
signs, 100
size perception, 96–98
stages of vision/perception, 99
Stiles Crawford effect, 79, 80
theory of vision, 96
unconscious inference and perception,
    98–101
visual acuity, 101–103
visual process, 72
visual psychophysics and, 67
Physiological optics, 63–64; *see also*
    Physiological optics
Pinhole camera, 161
Plasma-based sources, 169, 170
Plasma plume, 145
Plasmons, 139
p–n diode, 159
Post-mediaeval western culture, 123, 129
appropriation of Ibn-al-Haytham's work, 125
art, 126
Euclidean continuum, 127
literature, 125–126
optics, 125
religion, 128–129
theories of vision, 124–125
Power plant, first central, 183
Priming, 100
Protestant Reformation, 128–129
Psychophysics, 67
Ptolemaic assumptions, 59
Ptolemy, 57
assumptions, 59
hypotheses, 60–61
moon illusion, 89–90
moon model, 59
Pump-probe method, 169
Punctiform analysis, 79

R

R&D, *see* Research and development
Rationalist theology, 17–18

Rāzī, 19
Reasons of light to lights of reason, 23
age of the Enlightenment, 24
conception of light, 28
d'Alembert, Jean, 37
Descartes, René, 36, 37
Einstein's contributions, 30–36
Ibn al-Haytham's physical optics, 25–30
intromission of forms, 27
light as physical and rationality, 23–25
Lorentz's and Poincare's contributions, 31
Lorentz' transformations, 31
optics, 28
physics and rationality, 36–38
regional rationality, 25
significant moments in elaboration of science
    of light, 24
theories, 25
visual ray, 27, 28
Reflection, 45, 47, 48, 49
laws, 194
Reflexio, 45–47; *see also* Book of Optics,
    The
Refraction, 44, 45, 47, 48, 49, 195
law, 136, 197
Regional rationality, 25
Regional science centres, 211; *see also*
    International Science Centres in Arab
    Countries
ICTP, 212–213
Reject option decision rule, 96
Remote laser induced breakdown spectra, 146
Remote micro-imager (RMI), 144, 145
Research and development (R&D), 208
Research and jobs, 140
Risner, Friedrich, 43; *see also Book of Optics*,
    The
Ritter, Johann Wilhelm, 165
silver chloride and colour spectrum,
    165–166
RMI, *see* Remote micro-imager

S

Saccades, 81
Saint-Joseph University of Beirut (USJ), 243
Satyendra Nath Bose, 35
Scanning electron microscope (SEM),
    154, 156
School of Edessa, 230
Science
history of Arabic, 26
luminaries of, 64
and nature, 216
of optics, 29
Science and Technology National Policy
    (STNP), 232

Scientifically aware person, 235
Scientific awareness outreach programme, 235;
        see also Scientific translation
    acquisition of printed materials, 236–237
    collaborating institutions, 236
    database of scientific materials, 238
    focus groups, 235–236
    future scientist project, 237
    messages, 238
    multimedia, 237
    outreach projects, 236
    scientifically aware person, 235
    support of collaboration institutions to target
        audience, 237
Scientific centres, 216
Scientific method of enquiry, 160–161
Scientific theory, 62
Scientific translation, 229; see also KACST;
        Scientific awareness outreach
        programme
    annual citation impact, 240
    foundation of House of Wisdom, 230
    KSA's National Strategy, 231
    lessons learnt, 231
    number of approved patents per year, 241
    patent ranking of Arabic and Islamic
        countries, 241
    product space, 232
    ranking of different countries in scientific
        publishing quality, 239
    research outputs in Middle East, 240
    School of Edessa, 230
    schools of knowledge and translation, 230–231
    science, technology and innovation KPI,
        239–240
    translation role in civilisations, 230
Self-initiated motions, 92
SEM, see Scanning electron microscope
SESAME, see Synchrotron-light for
        Experimental Science and
        Applications in the Middle East
Shii interlude, 20
Shiis, 20
Short-wavelength pulsed light sources, 166; see
        also X-ray free electron lasers
    femtosecond x-rays, 171, 172
    fluxes of ultrashort hard x-ray pulses, 174
    HHG sources, 169
    motion of atoms inside molecules, 167
    plasma-based sources, 169, 170
    synchrotrons, 169, 170–172
    ultrashort x-ray pulses, 169
Short wavelength radiation, 166
Shutter camera, 167
Signs, 100; see also Physiological optics
SISSA, see International School of Advanced
        Study

SLAC, see Stanford Linear Accelerator Center
of Snell-Descartes, 27
Snellius' law, 5
Snellius, Willebrord, 193
Snell's law of refraction, 155, 156, 193
Solar Energy programme, 241
Southall, James P C, 64
Space-time, 32; see also Einstein
Special Theory of Relativity, 32–33; see also
        Einstein
Spherical aberration, 8, 155, 157, 162
    in cameras and telescopes, 158
    effect of correcting, 157
    Hubble Space Telescope images, 158
Spiritual vision, 129
SR, see Synchrotron radiation
Stanford Linear Accelerator Center (SLAC), 224
Stereoscopic vision, 84
Stereo vision, 145
Stiles Crawford effect, 79, 80
STNP, see Science and Technology National
        Policy
Sunlight, 158
Sunnis, 20
Superlens, 139
Synchrotron, 169, 170–172, 210
Synchrotron light, 198
    sources, 206, 221, 224
Synchrotron-light for Experimental Science
        and Applications in the Middle East
        (SESAME), 205, 206, 211, 221
    CESSAMag, 227
    Ibn al-Haytham, 221
    inspiration for SESAME people, 222–223
    installation of SESAME's storage ring,
        227–228
    international support and training
        opportunities, 225
    members, 223
    SESAME building, 222
    synchrotron light sources, 224
    timeline of, 224–225
    UNESCO support, 226–227
Synchrotron-light for Experimental Science
        and Applications in the Middle East
        (SESAME), 198
Synchrotron radiation (SR), 210
System of Copernicus, 54

T

TEM, see Transmission electron microscope
Theology, 21
Theories of vision, 96, 124, 136
    categories, 124
    Ibn al-Haytham's experiments, 124
    intromission theory of vision, 124

the substances and the accidents, 18
The World Academy of Sciences (TWAS), 212
Third World Organization for Women in Science
        (TWOWS), 212
Translation, 229
Transmission electron microscope (TEM),
        154, 156
Treasure of Optics, 43
Turnbull, 144
TWAS, *see* The World Academy of Sciences
TWOWS, *see* Third World Organization for
        Women in Science

U

Ultrafast optical spectroscopy, 169
Ultrafast structural dynamics with x-rays, 169
Ultrashort hard x-ray pulses, 174
Ultrashort x-ray pulses, 169
Ultraviolet radiation, 166
Unconscious inferences, 98–101; *see also*
        Physiological optics
UNESCO (United Nations Educational, Scientific
        and Cultural Organization), 225, 226
Universalisation of knowledge, 54
UPNEDA, *see* Uttar Pradesh New and
        Renewable Energy Development
        Agency
USJ, *see* Saint-Joseph University of Beirut
Uttar Pradesh New and Renewable Energy
        Development Agency (UPNEDA), 191

V

Vacuum ultraviolet (VUV), 169

Verbal communication, 229
Veselago, Victor, 139
Vision, 17; *see also* Physiological optics
        stages of vision/perception, 99
Visionary process, 17
Visual acuity, 101–103; *see also* Physiological
        optics
Visual ray, 27, 28
VUV, *see* Vacuum ultraviolet

W

Wavelength Division Multiplexing (WDM), 139
WDM, *see* Wavelength Division Multiplexing

X

XAFS, *see* X-ray absorption fine structure
XANES, *see* X-ray absorption near-edge
        structure
X-FELs, *see* X-ray free electron lasers
X-ray absorption fine structure (XAFS), 227
X-ray absorption near-edge structure
        (XANES), 166
X-ray absorption spectroscopy, 171
X-ray fluorescence (XRF), 227
X-ray free electron lasers (X-FELs), 169, 172
        seeded FEL, 173
X-ray spectroscopy, 166
XRF, *see* X-ray fluorescence

Z

Zewail, Ahmed, 208
Zone of uncertainty, 185

Printed and bound by CPI Group (UK) Ltd, Croydon, CR0 4YY

01/11/2024

01782619-0009